CYBER DEFENSE MATRIX

The Essential Guide to Navigating
the Cybersecurity Landscape

Sounil Yu

Clifford, Thanks for your contributions to our field! I hope you enjoy the book and that it helps you find new ways to put the matrix to use!

SY

Copyright © 2023 Sounil Yu

All rights reserved. No part of this publication may be reproduced, distributed, or transmitted in any form or by any means, including photocopying, recording, or other electric or mechanical methods, without the prior written permission of the author, except in the case of brief quotations embodied in critical reviews and certain other noncommercial uses permitted by copyright law. For permission requests, email the author at permissions@cyberdefensematrix.com.

ISBN: 979-8404393774
Imprint: JupiterOne Press

Publisher:
JupiterOne
2701 Aerial Center Parkway
Suite 120
Morrisville, NC 27560

Editor-in-Chief: Melissa Pereira
Copy Editors: J Daniel Janzen, Melissa Pereira, Gracie Yu, Valerie Zargarpur
Design: Dave Moy, Chum Wongrassamee
Formatting and Layout: Scott McFarlane

Soli Deo Gloria

To my patient wife Gracie and my children Caleb, Jason, Kristi, and Renee without whom this book would have either been completed a year earlier or perhaps never at all. (More likely the latter.)

To my patient friends and fellow practitioners who have been eagerly anticipating this book and encouraging me to finish it, I hope it was worth the wait.

Contents

FOREWORD *by Dan Geer* I

FOREWORD *by Wendy Nather* II

PREFACE V

1: INTRODUCTION 1

2: TERMINOLOGY 11

3: MAPPING SECURITY TECHNOLOGIES AND CATEGORIES 25

4: SECURITY MEASUREMENTS 45

5: DEVELOPING A SECURITY ROADMAP USING THE STACK 57

6: IMPROVING SITUATIONAL AWARENESS 73

7: UNDERSTANDING SECURITY HANDOFFS 87

8: INVESTING AND RATIONALIZING TECHNOLOGIES 99

9: DEALING WITH THE LATEST SECURITY BUZZWORDS 113

10: CONCLUSION 131

ACKNOWLEDGEMENTS 134

ABOUT THE AUTHOR 136

Foreword

by Dan Geer

As we all know, achieving security is a combination of People, Process, and Technology — a mantra worth repeating not because you haven't heard it before and not because there is any reason to argue against the proposition. Rather it is because perhaps the most difficult of the three is adequate process. Technologies for security are many, overlapping, frequently mutating (of necessity), oftimes conflicting, and no one of them can or will reverse the inherent strategic advantage that the offense possesses. Meanwhile, people are in permanently short supply. That leaves process, specifically process to address the security problem with the sort of abstraction layer that can keep the boat off the reef while yet outrunning the pirates. Yes, cybersecurity is all about tradeoffs, and competently handling tradeoffs is only doable when you can characterize the tradeoff space in strategic, conceptual terms. That is where this book comes in; it presents a strategic abstraction in the form of a categorization discipline that is simple without being simplistic, and sufficiently malleable to outlive the technologies and people you have to organize. Cybersecurity is the most challenging intellectual profession on the planet; however you approach it, your methods, your process has to be reliably stable (and straightforward) in the face of technologic ferment and sentient opponents. Start here.

Dan Geer, ScD

> Dan Geer is a security researcher with a quantitative bent. His group at MIT produced Kerberos, and a number of startups later, he is still at it today as Chief Information Security Officer at In-Q-Tel. He writes a lot at every length, and sometimes it gets read. He is an electrical engineer, a statistician, and someone who thinks truth is best achieved by adversarial procedures.

Foreword

by Wendy Nather

When it comes to describing the cybersecurity industry, we need fewer nouns and more verbs.

Never mind where the technology plugs in or how it integrates; what does it actually do? You would be surprised at the contortions of language that try to make the offering seem different: cutting-edge, next-generation, exciting. For a security professional who needs to understand all the myriad functions, techniques, breadth and scope of every applicable control, the security industry resembles an elephant created jointly by Dennis Ritchie and Picasso.

As a former chief information security officer and industry analyst, I have tried to eat this elephant in many different ways. I have used chopsticks, fondue forks, and my bare hands; I have tried it with magic quadrants, waves, and subway maps; I have ground up venture capital pitch decks to make Security Helper and drowned live demos in ketchup. Even queso, that miraculous Tex-Mex elixir, does not seem to make the consumption any easier. But once in a great while, someone creates a tool that truly works to make sense of a complicated view rather than adding to the confusion, and you are holding one right now.

With Sounil Yu's Cyber Defense Matrix, not only does this tool have a blunt end and a sharp end (for very high-level strategic discussions as well as nit-picky dissections of technical functionality), but it can telescope to different dimensions to encompass who owns a function or asset, who is responsible for handoffs in a process flow, and where risks have externalities. The mutually exclusive aspect of the Cyber Defense Matrix forces you to cut through the wishes and aspirations of a vendor's product (or your own software) to determine WHAT IT DOES, TO WHAT. Does the product "help protect X by alerting you when something happens to X?" Nope, that does not count as protection; that is detection. I do not care if you call it a CASB or a turbo-encabulator with quantic IoT; the nouns will change every year, but the verbs will drift less.

But the best part of all about Sounil's framework is that once you read this book about its multiple capabilities, you will realize that under the covers, it is a good primer for anyone on cybersecurity in general. As you read the examples, the edge cases, and how to analyze any asset class or function, you see in a relatively short time how it all should fit together — and recognize when something does not fit. You are armed with a clear, incisive vocabulary for communicating your security strategy, even if you never actually fill out the matrix. (For example, check out Chapter 4 for a way to distinguish the relative maturity of your favorite metrics.) This book's usefulness reaches far beyond "a place for everything and everything in its place."

I continue to learn more about cybersecurity every time I study the Cyber Defense Matrix, and I hope you will too. It will help you stomach a truly challenging field and give you the strength to continue your daily mission of defending what matters.

Wendy Nather

> Wendy Nather leads the Advisory CISO team at Cisco. She was previously the Research Director at the Retail ISAC, and Research Director of the Information Security Practice at 451 Research. Wendy led IT security for the EMEA region of the investment banking division of Swiss Bank Corporation (now UBS), and served as CISO of the Texas Education Agency. She was inducted into the Infosecurity Europe Hall of Fame in 2021. Wendy serves on the advisory board for the RSA Conference and Sightline Security, and is a Senior Cybersecurity Fellow at the Robert Strauss Center for International Security and Law at the University of Texas at Austin.

Preface

In my prior role as the Chief Security Scientist at Bank of America, one of my responsibilities was to meet with security startups to understand what capabilities they had to offer and determine whether or not these were needed within our portfolio of security controls. The sheer number of security startups, the size and complexity of our environment, and the lack of a common framework to express needs and corresponding solutions all conspired to make this matching exercise a wickedly hard problem.

The Cyber Defense Matrix was born out of a desperate need to address this problem. I wanted a common framework against which cybersecurity solutions could be mapped so they could be described and organized in a reproducible and consistent fashion. Although the original intent of the matrix was to map cybersecurity vendors, I found that there is tremendous depth to this two-dimensional matrix. It is easy to be fooled by its simplicity; the basic structure provides a mental map, which is easy to grasp and memorize, but as you begin to master its usage, you will discover there are many different angles from which you can view the matrix. Each angle provides a new and unique viewpoint from which your understanding of the cybersecurity landscape will grow.

Be warned that the fixed structure of the Cyber Defense Matrix frequently poses problems for me, and you will likely struggle with it too. It requires a decent amount of self-discipline to stick with the rule to align capabilities to fit within one box and only one box. I constantly encounter products and marketing language that seem to defy placement into a single box. I often want to give in and allow a vendor product to sit in multiple boxes. But ultimately, that is a cop-out. Wrestling through these situations requires patience and perseverance, and the outcome is usually the discovery of another viewpoint that reveals a deeper and more thorough understanding of our space.

There is a lot of hidden potential in the Cyber Defense Matrix. The longer I study it, the more it reveals, and the less I realize I know. And so, I invite you to read about what I have discovered, and I am also excited to hear about what new discoveries you make as you put the Cyber Defense Matrix to use in your environment.

CHAPTER 1

Introduction

*Everything should be made as simple as possible,
but not simpler.*
— Albert Einstein

Being Lost in the Cybersecurity Landscape

Imagine going to a grocery store to buy ingredients for a meal you want to cook. But when you arrive, instead of orderly aisles of products, you are confronted with a mountain of food piled haphazardly in the middle of the store. Even if you know exactly what the ingredients of the meal are, getting everything you need is going to be extremely hard because there is no organizational system to help you find them or compare competing products. Worse still, if the store does not have one of the items on your list, you will have to hunt through the entire pile before you can be sure of this. What a nightmare!

The cybersecurity ecosystem is just like this hopelessly disorganized grocery store. Consider the exhibit hall at any major security conference. The cacophony of claims from vendors hawking their wares, the confusing language of their marketecture, and the lack of any semblance of organization (aside from booths arranged roughly from largest to smallest) offer us no help to understand what we need or where to find it. Add in the rapid, unpredictable changes in the security environment driven by new attack surfaces and attack vectors, the increasing sophistication of threat actors, and the deluge of new capabilities arriving daily in the marketplace, and understanding the cybersecurity landscape becomes daunting even for the experienced practitioner.

Because there is no consistent terminology to describe cybersecurity needs and capabilities, there is much confusion about what a lot of products actually do — and vendors exploit this. Instead of a clear explanation of what a product does, practitioners are bombarded with overused fashionable buzzwords — often stretched so far in their application as to be almost meaningless. Practitioners need to stop letting marketing pitches drive the terminology and focus instead on plain English descriptors that tell us what a product really does.

The Cyber Defense Matrix was originally created to address this problem. It is an easy-to-memorize mental model that helps us navigate the cybersecurity grocery store to quickly find the capabilities that we need; compare and contrast features of similar products; and spot obvious gaps and deficiencies in our security posture.

This book will help practitioners understand this framework and how it can be applied to organize technologies. But it will also dive into how this simple framework can be extended to organize and understand many other aspects of cybersecurity. Join me as we use the Cyber Defense Matrix to understand and navigate the cybersecurity landscape.

Understanding the Landscape through the Cyber Defense Matrix

A map is an essential tool to understand the physical landscape of an environment that we may want to navigate. The more clearly a map distinguishes between different types of terrain so that they do not overlap (mutually exclusive), and the more completely that it captures all the features of the environment (collectively exhaustive), the more useful it will be as a navigational tool. Frameworks that are mutually exclusive and collectively exhaustive (MECE) make for great maps. Taking a MECE approach also allows us to partition a larger problem into smaller problems to make it easier to understand.

The Cyber Defense Matrix is a MECE representation of the cybersecurity landscape. It combines the five distinct functions of the NIST Cybersecurity Framework (**IDENTIFY, PROTECT, DETECT, RESPOND,** and **RECOVER**)[1] with five

---

1 I will use all caps for these five functions to distinguish from when the same words are being used in their general or ordinary sense.

distinct assets classes (**DEVICES, NETWORKS, APPLICATIONS, DATA,** and **USERS**)[2] to arrive at Figure 1. Plotting every defensive function against every kind of asset that needs defending creates this simple grid and offers a powerful organizing tool to understand and categorize much of what we do in cybersecurity. The Cyber Defense Matrix gives us a high-level overview of the entire cybersecurity environment for an enterprise and lets us see where any given cybersecurity product would fit into that enterprise.

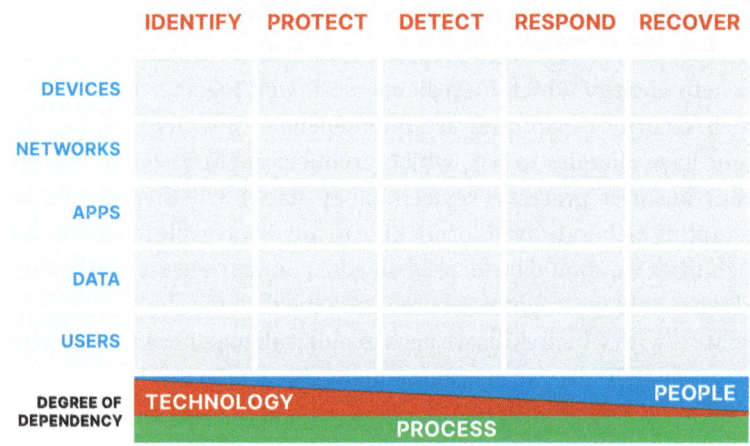

Figure 1: Cyber Defense Matrix

Most importantly, this MECE approach means the Cyber Defense Matrix can highlight which aisles and shelves in the cybersecurity grocery store are already fully saturated with an abundance of products, and which are empty and in need of attention. In other words, it can help find our blind spots, which is especially important in cybersecurity, where it is often something we missed or did not know about that can hurt us the most. The matrix takes what is implicit and sometimes forgotten, and makes it explicit to ensure that it is not forgotten.

The structure of the Cyber Defense Matrix serves as a forcing function to adhere to a consistent set of definitions and terminology. This consistency

---

2 Again, for consistency, I will use all caps when these terms are being used specifically in relation to their role in the matrix.

helps us cut through marketing language to quickly understand the basic function or purpose of specific security products. When we go into the vendor marketplace with the matrix in hand, we can quickly determine the core function of a given product and discern which products solve what problems.

In addition, the Cyber Defense Matrix helps us pick out the essential capabilities across the entire spectrum of options available to help secure our environment. As with food, we need to know what ingredients work well together in consideration of our diet. We also need to know if we have the necessary skills and staff to prepare the food. In a similar way, the matrix can help identify which ingredients work well together in a recipe (i.e., which security capabilities are interdependent); which ingredients we might have allergies to (i.e., which products might be overly disruptive to our business processes or technology stack); whether we should be attempting elaborate meals or sticking to microwaveable foods (i.e., which capabilities we should be focused on given our current state of maturity); and what help we might need preparing the dish (i.e., how many human operators with which skills we need to pull it all together.) These and other use cases will be described in subsequent chapters in this book.

It is important to understand that the Cyber Defense Matrix is not prescriptive. It provides a strategic 30,000-foot view of the entire range of options for security controls that an organization can implement to improve its security posture. Once we know at the strategic level what we need to do, the matrix can help us divide up the problem into discrete components and work it out at the tactical level. Given this purpose, the Cyber Defense Matrix may not be the most appropriate framework to use at the tactical level; other frameworks (e.g., MITRE's ATT&CK Framework) may be more suitable at lower levels of detail.

Now, just as a map does not tell us where to go, the Cyber Defense Matrix does not tell us what our security posture should be. However, once we know where we are and where we want to be, a map will help us get there. In the same way, the matrix can help guide us to improve our organization's security posture once we understand our current state and desired target state. The matrix helps us decide on a plan of action that is customized for our needs and the conditions of our enterprise.

Why "Cyber"?

"Cyber" is a much-abused term, overused to describe anything in our digital ecosystem. "Cyber" refers to everything and thus means nothing. Because the term lacks specificity, we lack a common understanding of what we mean by "cyber."

Instead, we often tend to define "cyber" from the narrow perspective of our own background and training. Those with backgrounds in system administration tend to think of cyber as being about endpoints and servers; those with software and product development backgrounds see cyber as software and application-centric; network administrators emphasize communication networks; those with a background in traditional information security think data is the most important focus for cyber; and those who have come out of personnel or physical security focus on people. Perhaps we collectively adopted the term "cyber" because we struggled to find the one word that encompassed all these different perspectives.

This ambiguity is carried over into how we define cybersecurity. Is cybersecurity about endpoint security? Or is it about application security? Or network and data security? How about insider threat? For all its flaws, "cyber" is the one word that seems to come closest to capturing the different types of assets in our digital ecosystem. But in using the word "cyber," we may quickly forget what each of these cyber assets are. To avoid leaving out an important cyber asset, we should be more explicit in defining the broader classes of cyber assets, which include devices, networks, applications, data, and users. Examples of each type of cyber asset are shown in Table 1.

Asset Class	Examples
Devices	Workstations, servers, phones, tablets, IoT, containers, hosts, compute, peripherals, storage devices, network devices, web cameras, infrastructure, etc. This class includes the operating system and firmware of these devices, as well as other software that is native or inherent to the device. Networking devices like switches and routers are included here because the devices themselves need to be considered separately from the communication paths they create.
Network	The communications channels, connections, and protocols that enable traffic to flow among devices and applications. Note that this does not refer to the actual infrastructure (e.g., routers, switches) but rather to the paths themselves and the protocols used in those paths. This means that areas such as DNS, BGP, and email filtering and web filtering also fall into this category. This class includes VPCs, VPNs, and CDNs.
Applications	Software code and applications on the devices, separate from the operating system/firmware. This class includes serverless functions, APIs, and microservices.
Data	The information residing on (data-at-rest), traveling through (data-in-motion), or processed by (data-in-use) the resources listed above. This class includes databases, S3 buckets, storage blobs, and files.
Users	The people using the resources listed above and their associated identities.

Table 1: Asset Classes and Examples

Not all of these asset classes may be equally important to a given enterprise. I often hear that data is the most important asset, or that it is all about the applications. There may be some truth to these aphorisms in a particular enterprise, but security practitioners that ignore or downplay whole classes of cyber assets do so at great peril. In the Cyber Defense Matrix, I try to treat all the asset classes as equally important, not because they necessarily are, but because they each offer attack surfaces that need to be properly examined and accounted for when considering the overall security environment.

Think about it this way: we can have the most secure device, running an application with no bugs, on a highly segmented network, with fully encrypted data. But if these assets are managed by a highly privileged administrator who clicks on every phishing email they get, then it does not matter how secure everything else is. The comprehensive character of the matrix — the fact that every function and subfunction must be considered for every asset class — makes it useful as a checklist in this regard.

When considering the varying types of assets, we also need to account for who owns those assets. In most cases, the assets that an enterprise cares about are those that are owned by the enterprise. However, we must also account for assets owned by other entities, such as vendors and third parties, customers, and employees. For some organizations, this list may also include assets owned by threat actors (as represented through threat intelligence). Asset ownership can add a third dimension to the two-dimensional matrix of asset classes and security functions. Table 2 lists example assets owned by other entities.

As we can see, there are a wide range of cyber assets. But why not just call them digital assets? In general, the prefix "cyber" is used in the context of security concerns about critical assets that warrant protection. The entities in each of these asset classes, including the broad range of different owners of these assets, are not just any digital assets, but rather assets that may be susceptible to attack.[3]

	Asset Owners			
	Vendors	**Customers**	**Employees**	**Threat Actors**
Devices	IaaS, EC2, ECS	Customer's computer	BYOD	Botnets
Networks	IaaS, CDNs	Customers' ISP	Residential ISPs	Bulletproof networks
Applications	SaaS, PaaS, Serverless	Customer's browser	BYOD apps	Malware
Data	S3 buckets, Block storage	Personally identifiable info	Personally identifiable info	Stolen info (e.g., credentials)
Users	Vendor admins, Vendor developers	Customers and their identity	Employees and their identity	Threat actor (e.g., Fancy Bear)

Table 2: Asset Owned by Other Entities

---

3 Herein lies a curious contradiction. In financial terms, assets are typically seen as resources that grow in value or help generate revenue. However, to the security practitioner, a cyber asset is one that introduces liabilities. These liabilities usually manifest in the form of new attack surfaces. Despite all the talk and excitement about digital transformation, it also translates into a rapid (and often unmanaged) proliferation of new attack surfaces (i.e., liabilities) that the security team must manage and mitigate.

Lastly, the term "cyber" is used for this framework to explicitly omit physical security considerations. This tightens the scope to areas that are less understood and need more attention. Although the practices of physical security are important and relevant for cybersecurity, these practices have been well understood for millennia and are covered thoroughly in other resources and frameworks on improving physical security.

Why "Defense"?

The Cyber Defense Matrix is a framework for the defense of an existing IT environment. Its first function is **IDENTIFY**, and more specifically, inventory. This implies that the asset to be defended already exists and must be enumerated. If an asset does not exist, there is nothing to defend. I am a great fan of the Cybersecurity and Infrastructure Security Agency's slogan: "Defend Today, Secure Tomorrow."[4] It makes clear that what we have right now must be defended, but that what we are building for the future should be made securely in the first place. In security engineering parlance, this is the difference between run-time controls (today) and build-time controls (tomorrow).

One encouraging development in cybersecurity is the emergence of "shift left" thinking and new processes that enable the creation of more secure cyber assets. This is an important goal; however, capturing these elements may be difficult within the Cyber Defense Matrix since it assumes that an asset already exists. Admittedly, this may be a flaw in the Cyber Defense Matrix, but if so, it would also be a flaw in the NIST Cybersecurity Framework itself.[5]

Dependency Curves

There is one more important piece to the Cyber Defense Matrix. At the bottom of the grid, there is a continuum that characterizes the degree of dependency on *TECHNOLOGY*, *PEOPLE*, and *PROCESS* as we progress through the five operational functions of the NIST Cybersecurity Framework.

---

4 https://www.cisa.gov/sites/default/files/publications/cisa_strategic_intent_s508c.pdf
5 Designing to ensure that assets are built without security flaws is a function of security architecture; this topic is notably absent from the NIST Cybersecurity Framework.

TECHNOLOGY plays a much greater role in **IDENTIFY** and **PROTECT**. As we move to **DETECT, RESPOND,** and **RECOVER,** our dependency on *TECHNOLOGY* diminishes and our dependency on *PEOPLE* grows. Throughout all five operational functions, there is a consistent level of dependency on *PROCESS*. This continuum helps us understand where we might have imbalances in our reliance on *PEOPLE, PROCESS,* and *TECHNOLOGY* when trying to tackle our cybersecurity challenges.

The shapes of these dependency curves often come under scrutiny by those who believe that *TECHNOLOGY* plays a larger role on the right side of the matrix. I welcome evidence to suggest that these curves are wrong, but thus far, the evidence that I have been presented with and the research that I have seen point in my favor.[6] Additional evidence for this shifting set of dependencies, and its subsequent implications, will be discussed in later chapters.

Use Cases for the Cyber Defense Matrix

Although my original use case for the Cyber Defense Matrix was to organize cybersecurity technologies, I have since discovered many additional use cases for this framework, which I will cover in subsequent chapters. These use cases include the following:

- Capturing and Organizing Measurements and Metrics
- Developing a Cybersecurity Roadmap
- Gaining Greater Situational and Structural Awareness
- Understanding Organizational Responsibilities and Handoffs
- Rationalizing Technologies and Finding Investment Opportunities
- Deciphering the Latest Industry Buzzwords

Before we jump into these use cases, it is important to make sure that we have our terminology correct. In the next chapter, I will explain the terminology that I will use throughout the rest of the book.

---

6 Strauch, Barry. (2017). Ironies of Automation: Still Unresolved After All These Years. IEEE Transactions on Human-Machine Systems. PP. 1-15. 10.1109/THMS.2017.2732506.

CHAPTER 2

Terminology

*There is no greater impediment to
the advancement of knowledge
than the ambiguity of words.*
— Thomas Reid, Scottish philosopher

Terminology Confusion

Words often are laden with baggage. What we think they mean might not always align with what they actually mean. This is particularly the case with the language we use in cybersecurity. The Cyber Defense Matrix uses the five functions laid out in the NIST Cybersecurity Framework (CSF): **IDENTIFY, PROTECT, DETECT, RESPOND,** and **RECOVER**. Unfortunately, our imprecise use of these terms taints our understanding of what each of these functions really means. These words are frequently used synonymously and interchangeably in marketing brochures, compliance requirements, and even in the NIST CSF itself. For example, the NIST CSF's definition of **DETECT** uses the word **IDENTIFY**: "Develop and implement appropriate activities to *identify* the occurrence of a cybersecurity event."[1] So, what is the difference between **IDENTIFY** and **DETECT**? Do we **IDENTIFY** events or **DETECT** events?

This confusion extends to foundational security concepts like vulnerabilities. Do we **IDENTIFY** vulnerabilities, or do we **DETECT** vulnerabilities? Again, the NIST CSF suggests that we **DETECT** vulnerabilities: "DE-CM-8: Vulnerability scans are performed."

---

[1] Framework for Improving Critical Infrastructure Cybersecurity, Version 1.1, April 16, 2018, https://doi.org/10.6028/NIST.CSWP.04162018, page 45.

How about the differences between **PROTECT** and **RESPOND**? When we want to remediate a discovered vulnerability or risk, is that a **PROTECT** action or a **RESPOND** action? If we were to read NIST's definition of Risk Management as "the process of identifying, assessing, and responding to risk,"[2] it sounds like it is a **RESPOND** action, but as we will soon see, that is not correct. Although they may seem interchangeable in our everyday vernacular, the actions are functionally very different.

Using physical analogies as an example, there is a major difference between knowing that a house is made of flammable wood (a vulnerability) and knowing that the house is on fire (an event resulting from an exploitation against that vulnerability). Likewise, there is a major difference between treating flammable wood to make it less flammable (mitigating a vulnerability) and putting out a fire (addressing an exploitation against a vulnerability).

The order in which the five functions are listed is significant, since each implies the existence of the prior function. Activities in **IDENTIFY** let us decide what to **PROTECT**. Similarly, activities in **DETECT** let us know what to **RESPOND** to, while **RESPOND** activities (and their success or failure) determine what **RECOVER** activities are required. We are often unable to **PROTECT** perfectly, so failures or bypasses in our defensive posture require timely notification and attention (i.e., **DETECT**) to such events.

Left and Right of Boom

To improve our understanding of these terms, it may be helpful to borrow the phrase "left of boom" and "right of boom," which are idioms originating from the U.S. military. In the original meaning, "boom" was the detonation of an improvised explosive device (IED). "Left of boom" activities focused on efforts to disrupt the ability of the attacker to create a boom. "Right of boom" activities focused on assessing and addressing the damage after the boom has occurred. Applying this concept to the five functions of the NIST CSF puts the functions of **IDENTIFY** and **PROTECT** on the "left of boom," or before a security event. **DETECT**, **RESPOND**, and **RECOVER** happen "right of boom," or after the event.

---

2 Ibid., page 46.

A fundamental benefit of the Cyber Defense Matrix structure is that it becomes a forcing function driving strict adherence to consistent functional definitions across all asset classes. In other words, all actions under **IDENTIFY** and **PROTECT** for all asset classes **(DEVICES, NETWORKS, APPLICATIONS, DATA,** and **USERS)** refer to "left of boom" activities. We cannot label a certain activity "**IDENTIFY**" for one asset class, and a similar activity "**DETECT**" for a different asset class. For example, if the discovery and inventory of **DATA** is properly aligned under **IDENTIFY**, the same function of discovering and inventorying **DEVICES** must also fall under **IDENTIFY**.

Table 3 provides a comparison of some of the key differences between left and right of boom activities.

Left of Boom	Right of Boom
• IDENTIFY, PROTECT	• DETECT, RESPOND, RECOVER
• Focuses on pre-event activities	• Focuses on post-event activities
• Associates with risk management	• Associates with incident management
• Aligns with security engineering	• Aligns with security operations
• Focuses on preventing intrusions	• Focuses on expelling intrusions
• Requires structural awareness	• Requires situational awareness
• Analyzing state	• Analyzing events & activity
• Inventorying assets	• Investigating state changes
• Discovers weaknesses	• Gathers evidence of exploitation against weaknesses

Table 3: Left and Right of Boom Activities

On the left side of boom, we need structural awareness of our environment. Structural awareness starts with an understanding of what assets you have, the state of those assets, and how the assets relate to one another. The state information includes how important the assets are, how they are configured, how they are exposed or weak, and how they are protected. The relational information characterizes how these assets are fitted together to make the whole system.

Boom occurs when a weakness is successfully exploited. Once that has happened, we are on the right of boom. We need situational awareness through an analysis of recent events and activities in our environment to understand if any of our assets have been compromised. But structural awareness is also vitally important here. If you are a firefighter about to run into a burning house, you will want structural awareness of the house, which includes answers to questions such as:

- What are the blueprints of the house?
- Is the house built to code?
- Are there locked doors that will impede our response?
- Where are the crown jewels that need rescue first?
- Are there toxic or dangerous materials?

Situational awareness and structural awareness complement each other to support incident response activities and will be covered in greater detail in Chapter 6.

What Does Each Function Mean?

The left/right of boom distinction is just one of the internal consistency checks the Cyber Defense Matrix uses to validate its definitions. The matrix uses the same words for the five functions as the NIST CSF. But owing to the internal consistency requirements of the matrix, the subfunctions and activities associated with each function in the matrix differ from those outlined in the NIST CSF.

The NIST CSF uses ordinary English language words like identify, protect, detect, respond, and recover. However, in their ordinary English meaning, these words are vague, overlapping, sometimes synonymous. It does not help that the framework authors repeatedly use the word "identify" to describe the function of **DETECT**, illustrating this confusion.

The NIST CSF authors never address these definitional issues and misalignment is the inevitable result. For example, there are a few instances where the same activity happens in two different functional categories. ID.RA-1 under **IDENTIFY** specifically mentions identifying vulnerabilities. But then we also have something about scanning for vulnerabilities in DE.CM-8 under **DETECT**. Are we not doing the same function there? Is not scanning for vulnerabilities in DE.CM-8 the same as identifying vulnerabilities in ID.RA-1? You can see how confusion may arise from labeling the same activity under two separate functional categories.

Understanding the differences between IDENTIFY and DETECT

When we consider the five functions of the NIST CSF with the left and right of boom dichotomy, it becomes clear how to distinguish between semantically similar terms such as **IDENTIFY** and **DETECT**. For example, if we consider the notion of a vulnerability as a structural weakness, then the discovery and enumeration of vulnerabilities is best aligned under **IDENTIFY**. Conversely, the exploitation of a vulnerability is best aligned under **DETECT**.

Suppose that a zero-day vulnerability is not discovered prior to it being exploited. Since this discovery occurs on the right of boom, should the discovery of this type of vulnerability be aligned under **DETECT**? If we see it that way, then the subsequent action, **RESPOND**, must also include patching that vulnerability. If we take this interpretation, we arrive at a situation where patching, normally a **PROTECT** activity, ends up being misconstrued as a **RESPOND** activity.[3] That cannot be right.

Instead, what actually happens is that we **IDENTIFY** a newly discovered vulnerability (that has already been exploited) and take actions to **PROTECT** ourselves from it so that future boom events can be avoided. At the same time, because that vulnerability has already been exploited, we must **DETECT** that exploitation and **RESPOND** to contain and eradicate any intruders. Patching the vulnerability (a **PROTECT** activity) will not expel intruders that are already present, but it can prevent future intrusions (i.e., it enables us to avoid getting to the right of boom this way again).

Understanding the differences between PROTECT and RESPOND

We want to bring this same consistency to the way we interpret the difference between **PROTECT** and **RESPOND**. During **PROTECT**, we address any deficiencies found through **IDENTIFY**. During **RESPOND**, we address any incidents found through **DETECT**. The correction of any deficiencies found during **DETECT** or **RESPOND** is still a **PROTECT** action since the intent of the correction is to avoid future exploitation against that deficiency.

---

3 This is a situation where the term remediation is also often unclear and misunderstood. Depending upon the context and how it is interpreted, remediation could refer to either a left of boom activity (remediation to address an open risk issue) or a right of boom activity (remediation to address an ongoing incident).

Returning to **IDENTIFY** and **PROTECT** for an exploited vulnerability does not imply a loop. Rather, as shown in Figure 2, it means that we are starting a new security activity thread which will systematically go through the logical order of the five functions, starting again with **IDENTIFY**.

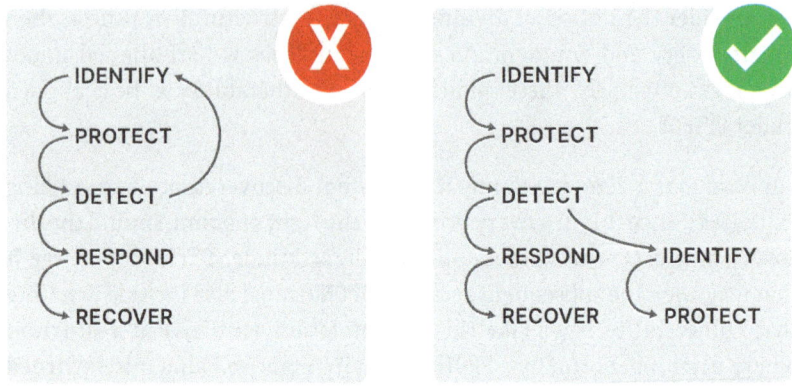

Figure 2: Sequencing of Activities Across the NIST CSF

Even though a previously unknown vulnerability may be revealed during **DETECT**-oriented activities (e.g., during an analysis of security events), it should still be considered an **IDENTIFY** function since it will trigger other related activities to determine where else is the vulnerability present (**IDENTIFY**) and if, where, and how it should be patched (**PROTECT**). In addition, for the existing incident, standard **RESPOND** and **RECOVER** activities would need to continue.

Understanding the differences between PROTECT and DETECT

There is further ambiguity around the function of **DETECT**. Most usages of the word "detect" in marketing materials are imprecise. The word "detect" can often be replaced with the word "logged," which is not actually performing the true function of **DETECT**. According to the NIST CSF, logging is a **PROTECT** function (PR.PT-1), but because logs are used for **DETECT**, the activity of logging is often mislabeled as **DETECT**.

When these logs are analyzed (DE.AE-2), correlated (DE.AE-3), and alerting threshold established (DE.AE-5), we are entering the realm of **DETECT**. Where it is nuanced and unclear is if we are filtering logs, particularly against a set of patterns that correspond to known attacks. Does the action of filtering fall under **PROTECT** or **DETECT**? What about other telemetry that we gather within our environment? How does that factor in? To help understand these differences, I offer a definition for these and other related terms that I will try to use consistently throughout the book:

1. *Telemetry* comes from instrumentation in **IDENTIFY** functions and entails information on the state of the asset. This Telemetry provides structural awareness and includes information such as the asset's configuration and vulnerabilities. Many **DETECT** tools require this Telemetry (which often leads to the commonly experienced "Yet Another Agent Problem").

2. *Logging* comes from **PROTECT** functions and captures information on Events, which are interactions with the asset and changes to the state of the asset.

3. *Filtering* is implemented on Telemetry and Logging to avoid information overload. Filtering is usually done in bulk and happens left of boom.

4. *Rules* can trigger against undesired state conditions, unexpected state changes, matches against a string value, volume thresholds, and many other parameters found in either Filtered data or in raw Telemetry and Logs. There are subtle nuances regarding the placement of Rules on the left of boom or right of boom. The point of making this distinction is to ensure that we have a clear understanding of expectations when we engineer Protection Rules versus Detection Rules. The degree of dependency curves on the Cyber Defense Matrix indicate that **PROTECT** activities should rely more on *TECHNOLOGY* whereas **DETECT** activities shift that reliance more towards *PEOPLE* (i.e., security analysts), and this applies to Rules as well.

5. *Protection Rules* are generally well defined (i.e., false positives are rare) with a highly deterministic course of action (i.e., all possible outcomes are known). Preventative action based on Protection Rules is generally taken left of boom leveraging technology and requiring minimal, if any, human intervention (e.g., antivirus match on known malware or intrusion prevention system match on known malicious traffic).

6. *Detection Rules* for when we wish a human analyst to be Alerted start moving us to the right of boom to **DETECT**. If a Detection Rule generates Alerts that are not meant for human consumption, but are sent to humans anyway, this can result in a tremendous amount of noise

and frustration for security analysts. Therefore, Detection Rules should trigger automated Enrichment so that analysts have as much contextual and environmental awareness (see Chapter 6 for the meaning of those terms) as possible to make well-informed decisions on whether or not to move to the next stage of **RESPOND**. Once this Enrichment has occurred to support analyst decision-making, it can be turned into an Alert.

7. *Alerting* is based on when a single or combined set of Rules are triggered. As a general best practice, a triggered Rule should be followed with as much automation as possible before Alerting because unenriched Alerts waste precious cycles of human analysts who are monitoring for Alerts.

Despite all the automation that we might have to support our **DETECT** functions, human analysts are central to the function of **DETECT**. The **DETECT** function should drive toward an incident management decision on whether or not to **RESPOND**.[4] Simple logging of events does not drive that outcome and should not be included as a part of a **DETECT** function; thus logging remains in **PROTECT**.

What Is an Asset?

When it comes to the five asset classes of the Cyber Defense Matrix (**DEVICES, NETWORKS, APPLICATIONS, DATA, USERS**), there is no logical order comparable to the one we see with the NIST CSF functions. They are listed in the same order each time for the sake of consistency, but the order itself is not significant. What is significant is that the asset classes are intended to be mutually exclusive and collectively exhaustive (MECE) as mentioned in Chapter 1.

Because the Cyber Defense Matrix aims to be a comprehensive checklist, to ensure that no asset is overlooked, it is more explicit and specific on what those assets are: **DEVICES, APPLICATIONS, NETWORKS, DATA,** and **USERS**. The matrix takes that which is implicit and makes it explicit and forces us to apply all five NIST CSF functions to all five asset classes. This internal

---

4 This mirrors the interplay between IDENTIFY and PROTECT where the end result of an IDENTIFY function is a risk assessment, which is used to make a risk management decision on whether or not to PROTECT.

consistency of the matrix is what drives much of its utility as a tool for systematically understanding our security environment. In addition, the term "asset" is not left ambiguous. By being more specific and consistent in defining an asset, the Cyber Defense Matrix can ensure that all potential attack surfaces in an organization are considered.

In contrast, the NIST CSF is not clear in defining what is an asset. For example, in the function of **IDENTIFY**, the NIST CSF refers to "systems, people, assets, data, and capabilities."[5] Systems, people, and data roughly correspond to some of the five asset classes in the Cyber Defense Matrix, but what exactly is an "asset" in this context? Later in the NIST CSF, we see asset management as a category of **IDENTIFY** (ID.AM).[6] There are **IDENTIFY** subcategories for **DEVICES** (ID.AM.1), **APPLICATIONS** (ID.AM.2), and **NETWORKS** (ID.AM.3). ID.AM.5 touches upon **DATA**, but only in the context of prioritization. And there is nothing at all for **USERS**, except for identifying their cybersecurity roles (ID.AM.6). Practitioners know that understanding who our **USERS** are is a critical part of **IDENTIFY**, yet the NIST CSF does not explicitly include it. A reference to the **USERS** asset class does eventually appear under the function of **PROTECT** in the category of Identity Management, Authentication, and Access Control (PR.AC), which lumps them together with systems that lock doors.

Gaps continue to appear as we move into **PROTECT**. The NIST CSF defines **PROTECT** as "Develop and implement appropriate safeguards to ensure delivery of critical services."[7] The notion of an asset is even more vague here than in **IDENTIFY**. What specific asset or classes of assets are being referenced? The NIST CSF does include categories like Awareness and Training (PR.AT), which correspond to **USERS-PROTECT**, and Data Security (PR.DS), which corresponds to **DATA-PROTECT**. But where are the categories that **PROTECT** our **NETWORKS**, **APPLICATIONS**, or **DEVICES**? There is a catch-all category of Protective Technology (PR.PT), which is defined as "ensuring the security of systems and assets." Perhaps this is where the remaining three asset classes are covered; however, the NIST CSF wording is still vague with respect to defining what exactly is an "asset."

---

5 Framework for Improving Critical Infrastructure Cybersecurity, Version 1.1, April 16, 2018, https://doi.org/10.6028/NIST.CSWP.04162018, page 7.
6 Ibid., page 24.
7 Ibid., p 7.

Likewise, in **DETECT**, specifically DE.CM-7, only four asset classes are covered: unauthorized personnel (**USERS**), connections (**NETWORKS**), devices, and software (**APPLICATIONS**). There is no explicit mention of any **DETECT** functions that cover **DATA**, even though **DATA** assets are covered in **PROTECT** (PR.DS).

Once we get into **RESPOND** and **RECOVER**, the NIST CSF ceases to explicitly call out any particular type of asset. Even being as generous as we can with these definitions, it is clear that these inconsistencies in the NIST CSF definitions leave too much room for gaps, as seen in Figure 3.

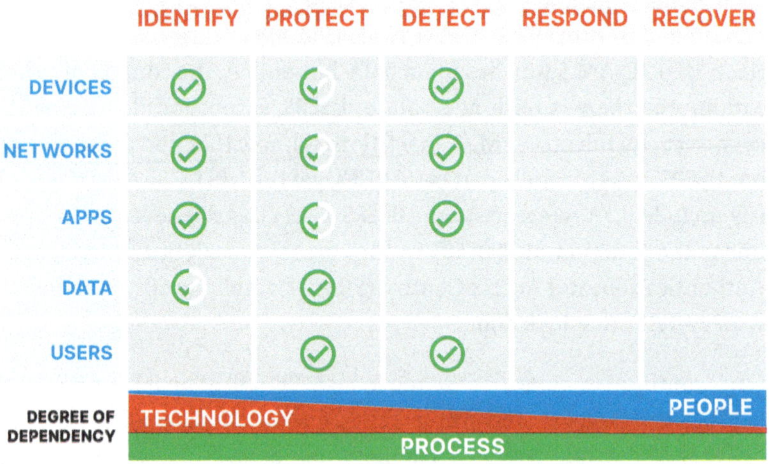

	IDENTIFY	PROTECT	DETECT	RESPOND	RECOVER
DEVICES	✓	✓	✓		
NETWORKS	✓	✓	✓		
APPS	✓	✓	✓		
DATA	✓	✓			
USERS		✓	✓		

DEGREE OF DEPENDENCY: TECHNOLOGY — PROCESS — PEOPLE

Figure 3: Asset Class Coverage in the NIST Cybersecurity Framework

These inconsistencies and vagaries in the NIST CSF make it much more likely that we will unwittingly leave holes in the defense of our assets. By being specific in defining what is an asset, and consistent in applying each of the NIST CSF functions to them, the Cyber Defense Matrix ensures that the full range of people, process, and technology capabilities can be properly mapped to each asset class. Through this mapping, we can then see gaps in our security posture across our whole environment.

Clarifying Ambiguities Among Asset Types

The terms that describe assets seem deceptively simple; however, they too can lack precision because the same terms can mean different things to different people. This can make it tricky to map items onto the Cyber Defense Matrix correctly and consistently.

This is exacerbated by the growing convergence and complexity of today's IT environments. In an era of cloud services, outsourcing, and hybrid infrastructure, it may not always be clear how some of the five asset classes are MECE. For example, what asset class is "cloud?" Does the term "infrastructure" refer to **DEVICES** (i.e., servers) or **NETWORKS**? Does firmware align with **DEVICES** or is it part of **APPLICATIONS**?

The Cyber Defense Matrix is not perfect. It is a work in progress, so let us start by laying out a gray area — a terminology issue I have wrestled with in developing the matrix.

Applications versus Devices

One of the areas that I have struggled with in developing the matrix is the question of where to draw the line between **APPLICATIONS** and **DEVICES**. At first glance, this might seem a simple question. **DEVICES** are hardware and **APPLICATIONS** are software, right? Well, yes, up to a point. But what about firmware? What about operating system software? Or the software that we write? When it comes to security controls, do we treat software that we build differently from the software that we buy?

Remember that the matrix is intended as a tool for practitioners to group similar items together. As a practical matter, when using the matrix, it makes more sense to treat the operating system as part of the **DEVICE** it is running on. When we look for vulnerabilities on a machine (**DEVICE-IDENTIFY**) we look at the operating system. If the operating system is vulnerable and gets exploited, it is the **DEVICE** that is at risk. To fix the vulnerability (**PROTECT**), we patch the operating system on the **DEVICE**.

Operating systems and firmware are at one end of the software spectrum. At the other end are our own in-house built apps, for which we own and develop the source code. In between is commodity software like email clients, web browsers, or tool suites like Microsoft Office.

The **APPLICATION** vs. **DEVICE** distinction is one place where the bright lines of the matrix can actually mask something of a gray area. There is a spectrum of different types of software. One good test to make sure we are mapping asset classes correctly is to move to the next function and see if the asset class remains the same, as I did when thinking about operating systems just now.

Using this test, we generally find that enterprises deal with commodity applications in the same way they deal with an operating system. We find **(IDENTIFY)** and remediate **(PROTECT)** vulnerabilities on the **DEVICE**. Even if the application is open source, and we have access to the source code, we will still deal with vulnerabilities the same way we do for commercial software. We look for them by scanning software on the **DEVICE**, and then fix them by applying a patch from the developer on the **DEVICE** where the program is loaded.

By contrast, when dealing with in-house built applications, vulnerabilities are found **(IDENTIFY)** by testing the source code itself with Static or Dynamic Application Security Testing. Any vulnerabilities we find are typically fixed in a development environment, not on the **DEVICE**. If we are unable to fix the application directly, we can use tools like a Web Application Firewall (WAF) or Run-time Application Self Protection (RASP) to offer another layer of **PROTECT** capabilities. The WAF is an indirect fix for a vulnerability. It does not cure the flaw, but it **PROTECTS** the flawed application.

This means that for in-house built applications, the capabilities to perform the functions of **IDENTIFY** and **PROTECT** are very different from what is needed for commodity software. Because of this, the Cyber Defense Matrix defines **APPLICATIONS** as only those programs that the enterprise has created and/or for which it builds and maintains the source code. Commodity applications, like operating systems and firmware, are considered part of the **DEVICE** asset class.

These definitions are admittedly somewhat counterintuitive, but remember that the matrix is designed for practitioners to find like-for-like items grouped together. If we have a mismatched mapping, then we might bring the wrong capabilities to the fight or fight to secure the wrong things. For example, in most enterprises, the in-house software included in the **APPLICATIONS** asset class comprises the crown jewel of the organization's intellectual property. Defending it is a key task for security practitioners. Drawing the line in this fashion ensures that the defensive functions we perform on the **APPLICATIONS** asset class are focused on those crown jewels. If the **APPLICATIONS** class includes commodity software, **APPLICATIONS-IDENTIFY** and **APPLICATIONS**-PROTECT will include a lot of technologies that do not help us defend our in-house software. Worse, we can end up with a gap in our security posture because irrelevant capabilities may obscure the fact that we have a gap there.

Users versus Identity

The term "Identity" is most often directly associated with the **USER** asset class. In some cases, I have seen the **USER** asset class replaced by the term "Identity." The main misconception here is that "Identity" is synonymous and unique to the **USER** asset class. However, every asset has an identity. For example, **DEVICES** have device certificates, **APPLICATIONS** have TLS/SSL certificates, **NETWORKS** have IP addresses, **DATA** has hashes and other metadata. All these are identity attributes that are distinct to each asset class. Identity management is a significant concern that deserves special attention and many of the shortfalls may be found in the **USER** asset class, but the challenges of identity management are not isolated to just the **USER** asset class.

Enforcing Functional Consistency

One of the Cyber Defense Matrix's most useful features is the internal consistency it enforces. Within a given function (e.g., **IDENTIFY**), the pattern of activities for each subfunction must repeat for each asset class. For instance, one of the first activities within **IDENTIFY** is the subfunction of inventory. This is often also expressed as asset visibility, and it precedes virtually all subsequent security activities. We must know about the existence of an asset and its attributes, especially those attributes that are uniquely associated with the asset (i.e., its identity), before we can proceed.

This subfunction of inventory applies for every asset class, but the terminology that we normally use may slightly differ for each class. The term "inventory" may not even be used; a **DATA** inventory might be better known as a data catalog, and a **USERS** inventory might be called a people or role directory. Regardless of the terms that are specific to a particular asset class, we should expect to see some comparable form of that activity within the **IDENTIFY** function for each asset class.

After inventorying, usually we want to prioritize or classify the asset (i.e., understand impact); examine its attack surfaces for weaknesses (i.e., understand vulnerabilities); and model the attacks that may come against it (i.e., understand threats). These are three key components to perform a proper risk assessment. In essence, for each asset class, we are asking the following four questions:

1. Do we have something (inventory)...
2. That we care about (impact)...
3. That has weaknesses (vulnerabilities)...
4. That someone is after (threats)?

If the answer is no, there is no need to progress further. But if the answer is yes, then we must make a risk management decision as to whether to move to **PROTECT** the asset under evaluation. In this fashion, the Cyber Defense Matrix, as it moves from function to function across each asset class, provides a consistency check to ensure that all subfunctions have been considered. For every function and for each asset, we can look forward and backward across the subfunctions, and up and down across the assets, to check that what we are doing is consistent.

Based on our risk tolerance, not every asset class might require the execution of every subfunction (particularly those under **PROTECT**) under every circumstance. But for each asset class, all subfunctions should at least be considered. We will have to ask ourselves why are we not doing subfunctions #1, #2, #3, and #4 for this asset class. We may choose to ignore a box or risk-accept the absence of a control, but the Cyber Defense Matrix will not let us forget about it.

This internal consistency check is a great illustration of the value of the matrix in providing a comprehensive view of our security environment, for it forces us to consider each step in turn. Every security activity has to go through that five-function cycle. And the cycle has to be repeated — even if some functions are modified or rejected altogether in some cases — for each of the five asset classes.

CHAPTER 3

Mapping Security Technologies and Categories

> *The true beginning of scientific activity consists ... in describing phenomena and then in proceeding to group, classify and correlate them.*
>
> — Sigmund Freud

Bringing Order to Marketing Chaos

In this chapter, we will consider the first use case for the matrix, and the one for which it was originally designed: mapping security technologies. By figuring out where security technologies and products sit on the Cyber Defense Matrix, we can ensure that we are defending our key assets, avoiding duplication or overlap, and identifying possible gaps or blind spots where we lack needed coverage.

This may seem a simple exercise, but it can be wickedly hard. Thanks to the vague, exaggerated, or otherwise misleading claims generally made in marketing literature, it is often difficult to figure out what a cybersecurity product actually does. A healthy dose of skepticism can help us boil away the marketing froth, but that is only the first step. We still need to replace it with more useful plain-English descriptors — and figure out where the product fits in our security environment.

Other approaches to this problem, such as the ones taken by Optiv[1] and Momentum,[2] use a taxonomy of some kind. However, this merely produces long lists of product categories and vendor logos. They are not in any particular order and, despite their length, there is no way to tell for sure whether the lists are actually exhaustive.

Some refer to themselves as a periodic table of cybersecurity elements. While that usually results in a pretty table, it is often misleading. In chemistry, each column of the periodic table contains elements that predictably behave in similar ways. For this reason, a true periodic table can reveal the nature of elements yet to be discovered. However, none of the periodic tables offered by various cybersecurity vendors and resellers exhibit this predictive power.

In contrast, the structure of the Cyber Defense Matrix provides a more methodical and consistent approach to organize these cybersecurity elements. By mapping products to the appropriate boxes on the matrix, we can ensure that we have the security capabilities we need in every area we need them.

The labels for the Cyber Defense Matrix boxes (e.g., **DEVICE-PROTECT**) are much less exciting and glamorous-sounding than the categories vendors typically use to describe their products (e.g., "autonomic vulnerability remediation"), just as "apples" is less appealing than "golden delicious" or "honeycrisp." But they are also less distracting. They are keyed to the basic technology categories that most practitioners are familiar with, not the frothy titles used to distinguish one product from all its competitors in a crowded and noisy market.

Since vendors come and go (and keep coming out with new or allegedly improved products while they are around), our focus here will be on mapping product categories — in other words, on providing rules of thumb to allocate capabilities to a particular box in the matrix. Specific mappings for contemporary vendor offerings can be found on the companion website: https://cyberdefensematrix.com.

---

1 https://www.optiv.com/navigating-security-landscape-guide-technologies-and-providers
2 https://momentumcyber.com/docs/CYBERscape.pdf

Some Rules of Thumb

Let us start with some broad rules of thumb which will help us map security technologies correctly.

For **IDENTIFY**, **PROTECT**, and **RECOVER**, the asset class mapped should be the one subject to the function, the first-order asset that is being identified, protected, or recovered, regardless of where the function actually operates. A common error in mapping occurs when a capability is mapped to where it lives versus what it secures. For example, a web application firewall lives on the network, but it is protecting applications, so the appropriate mapping is **APPLICATION-PROTECT**.

For **DETECT**, the asset class mapped should be chosen based on the use case rather than on where the telemetry used in the detection originates. For example, an insider threat tool may leverage telemetry from **DEVICES** and **NETWORKS**, but the use case is specific to the **USER** class of assets, so the appropriate mapping would be **USER-DETECT**.

For **RESPOND**, the asset class mapped should be based on the asset that is being responded to or investigated. For example, a forensic tool that digs through packet captures would be mapped to **NETWORK-RESPOND**. If it is a forensic tool that digs through a hard drive, it would be **DEVICE-RESPOND**. Forensic reconstruction of queries that were run against a database would be **DATA-RESPOND**. For security orchestration and response (SOAR) tools, the mapping is dependent upon the modules or connectors that enable response actions against a particular asset class.

One helpful trick to double check that a security technology is being mapped to its correct asset class and function is through pattern matching. There are three primary ways that the Cyber Defense Matrix facilitates pattern matching when mapping security technologies.
  1. Asset consistency patterns
  2. Functional consistency patterns
  3. First-order vs. second-order patterns

Asset consistency patterns. Once a technology has been mapped to a particular asset class, we can move forward to the next function in the matrix and confirm if related technologies still apply to the same asset class. For example, if a product finds vulnerabilities in applications (**APPLICATION-IDENTIFY**), then we should expect something in the next function (**PROTECT**) that can address those vulnerabilities, such as runtime application self protection (RASP). If we do not, that is a sign that we need to revisit our mapping decision.

Take vulnerability enumeration for commercial software such as Adobe Acrobat, for example. Our first instinct might be to place this product in **APPLICATION-IDENTIFY**. If the pattern holds, we will expect the technologies for patching that software to be found in **APPLICATION-PROTECT** — but they are not. Instead, they are found in **DEVICE-PROTECT**. This suggests that the vulnerability enumeration tools for commercial software should actually be in **DEVICE-IDENTIFY**.[3]

Functional consistency patterns. Once we map a technology to a particular function, we can check to see if the subfunctions performed are consistent when applied to different asset classes. For example, given that a network firewall is a form of access control for a **NETWORK**, it would seem to map to **NETWORK-PROTECT**. To check that **PROTECT** is the correct function for access controls, we can verify if the subfunction of using access controls (usually through deny lists or allow lists) to prohibit or constrain access to some asset continues to make sense when aligned under **PROTECT** for other asset classes. When this subfunction is applied to **DEVICES**, the deny lists can be virus signatures (e.g., antivirus), and the allow lists support the capability typically referred to as application control.[4] When this subfunction is applied to **DATA**, the deny lists can be personal identifiable information (PII) or credit card numbers (e.g., data loss prevention), and the allow lists can specify data that usually gets flagged as false positives. In a similar fashion, we can continue to apply this consistency check for all other asset classes to determine that the subfunction of access controls

---

3 This points to the general confusion that we have when we talk about securing applications. At a high level, we are trying to secure applications that were built either by us or by someone else. If they were built by us, then the security capabilities will tend to be found in the APPLICATION asset class. On the other hand, if they were built by others (and bought by us), then the security capabilities will tend to be found in the DEVICE asset class instead.

4 As mentioned in the footnote above, the term "application" here generally refers to black box software that organizations buy. Therefore, "application control" maps to the DEVICE asset class.

using deny lists and allow lists falls best under **PROTECT**.

Data loss prevention (DLP) serves as a good example for a failed functional consistency pattern check because it is commonly miscategorized as **DATA-DETECT**. This is because DLP is often implemented in monitoring mode only, without blocking any content because it is too disruptive to normal business operations. However, DLP maps to **DATA-PROTECT** even if it is not actually preventing any data leakage. Regardless of whether one chooses to operate in 100% blocking mode, 0% blocking mode (i.e., monitoring mode), or somewhere in between, the mapping of the capability does not change because its core function has not changed. A DLP capability operating at 0% blocking is still mapped to **DATA-PROTECT**, and a firewall operating with an "any-any" rule (i.e., not blocking anything) is still mapped to **NETWORK-PROTECT**. The fact that one may choose to operate such capabilities without taking advantage of its core functionality does not change the mapping of that capability within the matrix.

First-order vs. second-order patterns. What does a firewall do? At the first-order level, its primary purpose is to **PROTECT** the **NETWORK**. However, I often get the answer that it also **PROTECTS DEVICES, DATA**, and all other sorts of assets. In addition, similar to the inconsistency mentioned above about DLP being a **DETECT** tool, I often hear that a firewall maps to **NETWORK-DETECT** because the telemetry and alerts from a firewall can be used to hunt for intrusions.

While this may be true as a second-order effect of the firewall, it is not as a first-order purpose. When mapping to the Cyber Defense Matrix, the capability should be placed against the asset and function that align best with the capability's primary, first-order asset and function. If it seems that a specific capability aligns against multiple assets or functions, we will likely find that we are mapping to its secondary purpose. It is easy to get fooled by marketing exaggeration. A common indicator that a product feature is achieved as a second-order effect are qualifying words such as "helps," "supports," and "enables." These words usually indicate that another separate capability is required in order to achieve the desired effect.

The problem with mapping to second-order effects is that it introduces too many unbounded possibilities. For example, we (and many vendors) could claim that many security capabilities **PROTECT DATA**, but most do that as a byproduct of protecting something else first. An endpoint protection platform (EPP) primarily maps to **PROTECT-DEVICE**, and as a secondary capability could map to **PROTECT-DATA**, but the second-order effect could also include **PROTECT-USER**, **PROTECT-NETWORK**, and everything else that the endpoint touches. Because the second-order effects are practically limitless, they should not be considered when mapping to the Cyber Defense Matrix.

Security Technology Categories Mapped

At this point, it might be useful to go through a list of various security technology categories and understand how I have mapped them. By explaining the thinking behind each of my choices, I hope to offer a deeper understanding of the structure and function of the Cyber Defense Matrix. It may be useful to pause and consider how you might map each of the categories below before proceeding to read my explanation of its mapping.

Email or web security gateway. Gateways serve as a means to **PROTECT**, so the alignment to the function is straightforward. However, which asset class does email and web security map to? Does it map to **DEVICE** since we commonly find email clients and web browsers on devices? Does it map to **USER** since these gateways help users avoid encountering malicious phishing emails and websites? Does it map to **NETWORK** since it operates on the network and controls the flow of specific types of traffic? Does it map to **DATA** since these help prevent phishing attacks which could lead to a data breach? It seems as though this category could map to multiple asset classes. When this happens, it is likely that we are including the second-order effects of the capability. When we consider the primary function of a gateway, it is to control access to a resource. For this category, the primary resource in question is the email client or web browser on a user's **DEVICE**, so the mapping for these two categories is **DEVICE-PROTECT**.

DDoS mitigation. DDoS attacks are generally intended to consume two types of resources: network bandwidth and application processing cycles. This corresponds to the **NETWORK** and **APPLICATION** asset classes respectively, and products in this category typically focus on only one or the other. These capabilities do not actually prevent a DDoS attack, but rather minimize the impact from such an attack. When a DDoS attack occurs, incident responders often enable DDoS mitigation services to dampen or eliminate the impact of the attack by filtering and/or sinkholing traffic, and as such, the mapping for this category is **NETWORK-RESPOND** or **APPLICATION-RESPOND**.

Some vendors may offer these services in an "always-on" mode, preemptively examining incoming requests, searching for previously identified patterns of a DDoS attack, and routing any such requests through the DDoS mitigation provider's own network infrastructure for network-centric DDoS attacks. This would suggest mapping to the function of **PROTECT**; however, these capabilities do not actually stop or prevent a DDoS attack. Rather, they provide a faster, more immediate *response* to a DDoS attack, similar to the way a sprinkler system in a building does not stop or prevent a fire, but helps reduce its impact.

Data loss prevention. Data loss prevention (DLP) is equivalent to a **DATA** firewall, which can be set up to prevent certain kinds of information from traversing specific boundaries in the enterprise. Although it has the capability to block data movement, DLP is often configured to simply monitor and log events instead due to the high number of false positives often observed by operators. Blocking and event logging is considered a **PROTECT** function; however, the fact that DLP is often configured with 0% blocking does not change it from being a **PROTECT** function. DLP controls are typically installed on the network, but they can also be **DEVICE**-centric (e.g, restrictions on whether data can be written to removable media) or **APPLICATION**-centric (e.g., SaaS DLP). However, this does not change the mapping because, as a **PROTECT** function, DLP is mapped to the asset class being protected — not where it operates. As such, the mapping for the category of DLP is **DATA-PROTECT**.

Data backup. At first glance, this might look like a **PROTECT** activity. Backing up data could be seen as protecting the availability of the data. Making the backup happens left of boom. However, it is not put to use until right of boom — after something bad has happened — which points to a right of boom mapping instead. We can also use pattern matching to see that data backup does not belong in **PROTECT**. Making a copy and storing it someplace for future retrieval does not fit the pattern of what the **PROTECT** function does in relation to any other asset class. Indeed, the closest analogue to a data backup — swapping in clean machines for infected ones — is considered a **RECOVER** activity, even though the machines will likely be purchased left of boom. In fact, if you think about it, a lot of activities to support **RECOVER** should be in place left of boom before any event has happened. We would not want to be doing our planning for business continuity or disaster recovery after a security event has happened.

Data discovery and classification. The asset class is clearly **DATA**, and the underlying activity is similar to inventorying, in that this capability is intended to generate a comprehensive data catalog. In some products, the discovery process also uncovers vulnerabilities. An example of such a vulnerability would be an open file share or a publicly accessible Amazon Simple Storage Service (S3) bucket. Sensitive data that is left unencrypted and exposed in a way that might allow access by unauthorized people would also be considered a vulnerability. How does one know that a particular piece of **DATA** is sensitive? The data owner would need to prioritize it, and one of the main means for doing this is through data classification tools. Some of these tools work in conjunction with data discovery tools, while others can operate as a stand-alone capability. Whether cataloging **DATA**, finding vulnerabilities in **DATA**, or classifying **DATA**, this capability aligns to **DATA-IDENTIFY**.

Data masking, encryption, tokenization, etc. These **DATA**-centric capabilities focus on limiting access to sensitive data, either by removing it or by converting it to other forms that it is not exposed during normal operations. Thus, this maps to **DATA-PROTECT**.

Application security testing. As the name suggests, this aligns to the **APPLICATION** asset class. One of the subfunctions of **IDENTIFY** is vulnerability identification, and that is what application security testing tools do — identify where we might have vulnerable code in our in-house built software. They do not perform the actual remediation of the vulnerability, which would be a **PROTECT** function. Thus, the mapping is **APPLICATION-IDENTIFY**.

Web application firewall. Web application firewalls (WAF) examine web application traffic and are set to block malicious web requests or alert according to a defined set of rules. Although WAFs live on the **NETWORK**, they shield web **APPLICATIONS** from attacks that attempt to exploit vulnerable application code. One could argue that, because a WAF shields against SQL injection attacks (which can expose sensitive information), WAFs align against the **DATA** asset class. However, this is a second-order effect. The primary effect is to shield the application itself from attack. In addition, some implement WAFs in monitoring mode, choosing not to implement some of the blocking features for fear of breaking their web applications. Regardless of whether a WAF is operating at 100% blocking mode, 0% blocking mode, or anywhere in between, it is still performing a **PROTECT** function. Thus, the mapping for the WAF category is **APPLICATION-PROTECT**.

Network intrusion detection system (IDS) vs. **network intrusion prevention system** (IPS). What differentiates an IDS from an IPS? An IPS, as the name suggests, prevents **NETWORK** intrusions by stopping known attacks before they progress further. This naturally aligns with the **PROTECT** function, so the IPS maps to **NETWORK-PROTECT**.

If an IPS is deployed leveraging 100% of its proactive blocking capabilities, it will likely break legitimate and critical business processes. One common way to avoid this is to deploy the IPS in monitoring mode (0% blocking) and then gradually turn up the dial, enabling more and more rules, signatures, and heuristics that block suspicious traffic until something important breaks. In monitoring mode, an IPS is basically like an IDS. The term "detection" in IDS may seem to suggest that an IDS maps to the function of **DETECT**; however, as with DLP and WAFs, the mapping of the capability does not change when we choose to operate it at a lower level of effectiveness.

As a rule, changing the way that technology is configured should not change its mapping. Regardless of whether the capability operates at 0% blocking, 100% blocking, or something in between, the activity of logging events is common throughout. Event logging is not sufficient for a capability to be mapped to **DETECT**. If that were the minimum qualification, then we could have virtually any capability that generates logs be labeled as **DETECT**.

To remain consistent with the structure of the Cyber Defense Matrix, an IDS, despite its name, should align as a **PROTECT** function, not a **DETECT** one. It simply logs network activities and generates events based on what it deems to be suspicious traffic. On closer inspection, most IDS events turn out to be false positives. The activity of a human actually monitoring those alerts and deciding what action (if any) to take is the essence of the true **DETECT** function, and that usually does not happen with an IDS, but rather with the next security capability below.

Security information and event management. SIEMs provide situational awareness by aggregating event telemetry from various log sources and correlating them together to help security analysts determine if an intrusion or compromise has occurred. This is a right of boom activity and aligns with the function of **DETECT**.[5]

The asset class mapping depends on our analytic use-case for the SIEM. If we are using a SIEM to consume IDS and firewall logs to help us discover an intrusion in our **NETWORK**, then the mapping is **NETWORK-DETECT**. Similarly, if we are ingesting Windows logs and anti-virus logs into a SIEM to find a compromised endpoint, then the mapping is **DEVICE-DETECT**. Consuming DLP logs to look for compromised data would map to **DATA-DETECT**. Processing WAF logs to look for a compromised web application would map to **APPLICATION-DETECT**.

Traditionally, these use cases were performed in isolation, leveraging narrow sets of information across a limited set of assets to discover intrusions. With the advancement of technology, we are seeing tools that can bring broader sets of information across the complete range of assets described by the Cyber Defense Matrix. This convergence is at the heart of the eXtended Detection and Response (XDR) category of security products, which I will describe in greater detail in Chapter 9.

Phishing simulations and tests. Just as we scan for vulnerabilities in other asset classes, we also want to routinely scan for vulnerabilities and potential exposures in our **USERS**. One common mechanism to conduct this "scan" is through a background check. Unfortunately, these are usually done only at the initial onboarding and are often not done continuously. Phishing tests would be equivalent to conducting regular, continuous vulnerability scans of **USERS**. By sending a properly calibrated

---

5 There is a degree of uncertainty at the heart of the DETECT function. If we could consistently determine that a defined set of events is always malicious or unwanted (i.e., DETECT), then it makes sense for us to proactively block it. Why would we allow a "boom" if we can prevent it (i.e., PROTECT) without causing any other business disruption? Many events analyzed by a SIEM start off in an uncertain state. Over time, we might gain a better understanding of how to interpret a given set of events, allowing us to confidently block the actions that led to the events without any unintended consequences. When this happens, the analysis and corresponding actions can move from being reactive (DETECT and RESPOND) to being proactive (PROTECT).

simulated phishing email, we can determine how susceptible a **USER** is to a phishing attack. Those **USERS** who are insufficiently skeptical about downloading unknown attachments or clicking links from suspicious senders are likely to be phished and should thus be considered more vulnerable. Vulnerability scanning is an **IDENTIFY** function, so this category maps to **USER-IDENTIFY**.[6]

Phishing awareness training. We often assume that if a user is vulnerable to phishing emails, we have to immediately fix that vulnerability. However, this is not how we treat other vulnerabilities in other asset classes. We typically conduct a risk assessment first. In the case of **USERS**, we should consider other threat and impact factors such as their role or level of access. If the combination of these factors results in a sufficiently high level of risk, then the user's vulnerability should be addressed. The vulnerability remediation often manifests in the form of security awareness training. In the same way that we patch vulnerable servers, we patch user vulnerabilities through training. Patching is a **PROTECT** function, so the mapping for this category is **USER-PROTECT**.

User behavior analytics. Insider threat is a common concern across many organizations. User behavior analytics (UBA) tools make it easier to discover this activity. UBA tools leverage two kinds of attributes associated with a **USER** to spot potentially malicious behavior: first, stateful attributes about the **USER** (e.g., age, job function, credit rating, etc.); and second, events and behavioral attributes surrounding the **USER** (e.g., time and place of **DEVICE** logins, type of activity on the **NETWORK**, **APPLICATIONS** access, etc.). In isolation and without proper context, these events are usually not

---

6 I often hear horror stories of companies that attempt to conduct a simulated phishing campaign only to suffer backlash from their users as they complain about how unfair a particular campaign is. Usually, this happens when the simulation is not properly calibrated to the skill level of the individual. The phishing campaigns that get the most complaints are those that employ more advanced techniques but are delivered to individuals whose skill level or role are not commensurate with the phishing test's difficulty level. This is the equivalent to giving a calculus test to a kindergartner and then punishing them for failing, a surefire recipe for complaints and tantrums. This calibration should also account for the role of the individual. For example, a system administrator of critical systems may warrant receiving more aggressive or sophisticated phishing tests. This pattern of calibration can be seen in other asset classes (e.g., rigorous scanning of Internet-facing servers) and should be adopted when scanning USERS for vulnerabilities.

sufficient on their own to deem particular **USER** behaviors malicious or even anomalous. However, with proper aggregation, correlation, and contextual analysis of these events, UBA tools can help determine if a **USER** is behaving as a compromised individual.[7] This analysis to determine if a compromise has occurred is a **DETECT** operation. As such, UBA maps to **USER-DETECT**.

A Note on the USER Asset Class

It goes without saying that **USERS** are different from the other four asset classes. While the parallel structure of the Cyber Defense Matrix is intended to show how various security functions seem to repeat themselves across all five asset classes, we cannot treat people as machines, like the way we treat devices or other technology. People have rights and dignity, and they have agency, too. They can choose their own path — which also means they have the capability to make well-intentioned mistakes. These differences suffuse everything about the way the Cyber Defense Matrix approaches the **USER** asset class, but they are especially important to take into account when considering **RESPOND** functions, e.g., how to deal with an insider threat once they are confirmed to be acting maliciously. For machines, a typical **RESPOND** action could include termination and aggressive forensic investigation. The expectations are completely different for **USERS** than for other asset classes because not only can people make mistakes, but they can also learn from them if counseled and trained properly in the aftermath.

Mapping Technologies to the Multi-Dimensional Matrix

Now I am going to look at a number of ways security technologies get mapped onto the three-dimensional matrix we discussed in Chapter 1, in which assets owned and/or controlled by other parties like vendors,

---

7 Account takeover may result in outside actors posing as employees but exhibiting the characteristics of an insider threat. Regardless of whether the actual employee is compromised or the employee's digital persona is compromised, the goal is to determine if we have a compromised USER in our midst.

customers, employees, or even threat actors each get their own layer of the matrix as shown in the following figures.

Threat Actor Assets

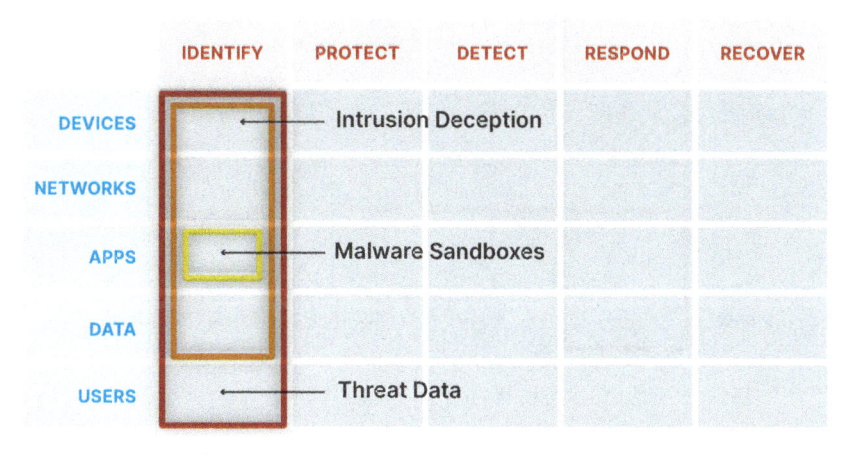

	IDENTIFY	PROTECT	DETECT	RESPOND	RECOVER
DEVICES	Intrusion Deception				
NETWORKS					
APPS	Malware Sandboxes				
DATA					
USERS	Threat Data				

Figure 4: Security Category Mapping for Threat Actor Owned Assets

Customer Assets

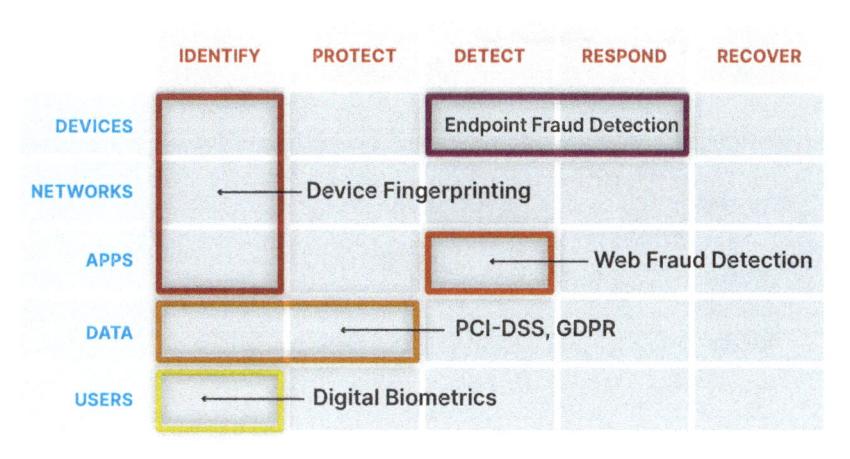

	IDENTIFY	PROTECT	DETECT	RESPOND	RECOVER
DEVICES			Endpoint Fraud Detection		
NETWORKS	Device Fingerprinting				
APPS			Web Fraud Detection		
DATA	PCI-DSS, GDPR				
USERS	Digital Biometrics				

Figure 5: Security Category Mapping for Customer Owned Assets

Vendor Assets

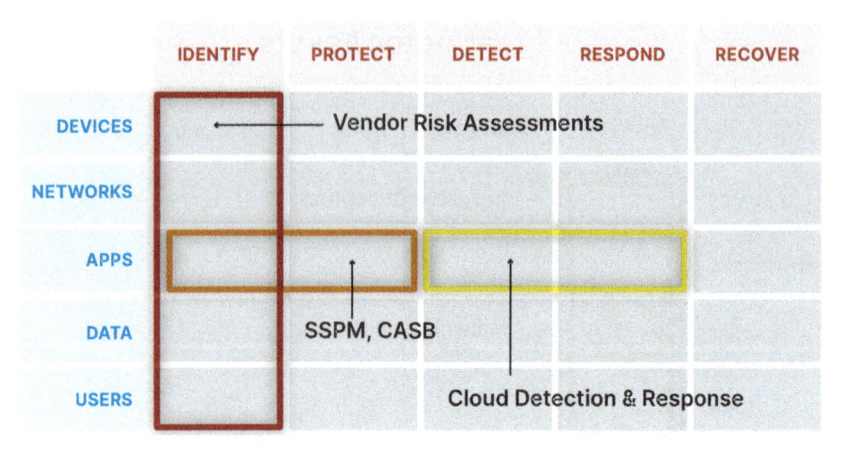

	IDENTIFY	PROTECT	DETECT	RESPOND	RECOVER
DEVICES	← Vendor Risk Assessments				
NETWORKS					
APPS					
DATA		SSPM, CASB			
USERS			Cloud Detection & Response		

Figure 6: Security Category Mapping for Vendor or Third Party Owned Assets

Employee Assets

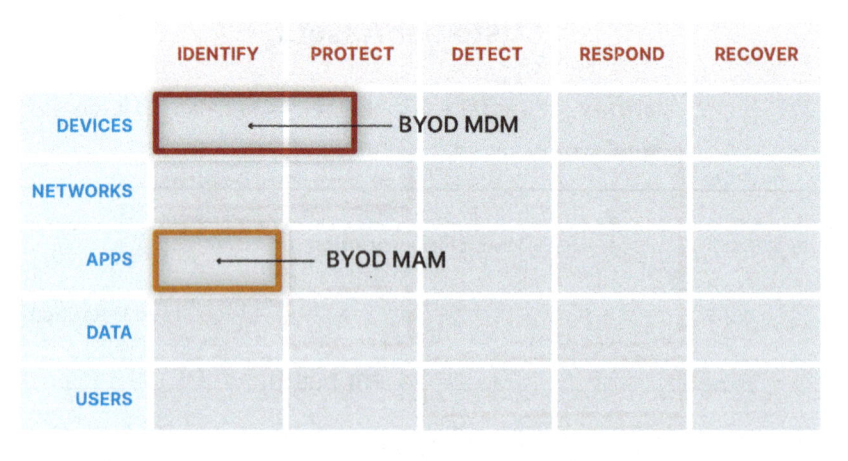

	IDENTIFY	PROTECT	DETECT	RESPOND	RECOVER
DEVICES		← BYOD MDM			
NETWORKS					
APPS		← BYOD MAM			
DATA					
USERS					

Figure 7: Security Category Mapping for Employee Owned Assets

Extending the matrix in this way also enables us to capture additional security capabilities without crowding out core capabilities that focus on securing assets owned directly by the enterprise. Each layer is aligned with a different asset owner, enabling a higher level of detail to capture capabilities that are distinct for different types of asset owners. Here are examples of how security capabilities map into the extended layers for assets owned by Threat Actors and Vendors.

Threat Actor Owned Assets

Threat Intelligence. The activity of threat intelligence is the gathering of information about threat actors. Threat actors also own **DEVICES, NETWORKS, APPLICATIONS,** and **DATA**. The threat actors themselves represent the **USERS** part of the matrix. Gathering details about their assets is similar to the activity of inventorying, which aligns to **IDENTIFY**. We can then think of threat intelligence as the activity of inventorying all of the threat actors' assets, including attributing who they are. Various threat intelligence providers specialize in one or more of the five asset classes. The following examples show how these threat intelligence capabilities can be mapped to specific threat actor assets and aligned against the Cyber Defense Matrix.

- Compromised hosts »→ **DEVICE-IDENTIFY**
- Malware »→ **APPLICATION-IDENTIFY**
- Bulletproof networks »→ **NETWORK-IDENTIFY**
- Stolen data »→ **DATA-IDENTIFY**
- Attacker attribution »→ **USER-IDENTIFY**

Deception Technologies. These technologies are designed to attract attackers, so they might seem to fall into the function of **DETECT**. But in fact, deception technologies work on the threat actor owned asset layer, and they are defined as part of the **IDENTIFY** function. Deception technologies provide information about threat actor assets, their targets (e.g., honey documents map to **DATA-IDENTIFY**), the machines they are using (**DEVICE-IDENTIFY**), and their tools (e.g., the malware and legitimate applications they are abusing, all map to **APPLICATION-IDENTIFY**). They can also tell us where threat actors are coming from (**NETWORK-IDENTIFY**). The mapping aligns under the function of **IDENTIFY**. The asset mapping is based on which type of attacker asset is being enumerated by the deception technology.

Sandbox technologies. These classify malware deployed by attackers and map to **APPLICATION-IDENTIFY** in the threat actor owned asset layer.

Note that with attacker assets, we have no desire to **PROTECT** their assets. We want to inventory them (**IDENTIFY**), and some law enforcement or intelligence agencies might want to know their true identity (**USER-IDENTIFY**) or the vulnerabilities that their hosts have (**DEVICE-IDENTIFY**), but there is no need to expend our energy to secure a threat actor's assets.

Vendor Owned Assets

Cloud access security broker. A cloud access security broker (CASB) appears in the vendor layer of the matrix because the assets being secured are owned by a third party. Typically, the owner of the asset would be a Software-as-a-Service (SaaS) provider. Although a few major SaaS providers offer enterprise-level security controls, most SaaS providers lack these features. So CASBs exist to wrap a layer around SaaS applications to enable enterprise security features such as application discovery, fine grained authorization to specific features, DLP, and logging. In a nutshell, CASBs attempt to extend enterprise-class security controls into a vendor-owned application. The mapping for CASBs is **APPLICATION-IDENTIFY** and **APPLICATION-PROTECT** in the vendor owned asset layer.

CASBs have generally failed to provide **DETECT** or **RESPOND** capabilities. Two new technologies purport to address these gaps: cloud detection and response, and extended detection and response (XDR). The mapping for these security capabilities are **APPLICATION-DETECT** and **APPLICATION-RESPOND** in the vendor owned asset layer.

Vendor risk assessment/third-party risk scoring. Companies offering vendor risk assessment and third party risk scoring gather information about a vendor's assets and perform a security assessment of those assets to understand the security posture of that vendor. These companies employ one of two methodologies for collecting this information: outside-in assessments and inside-out assessments. Outside-in assessments look at public-facing assets. These assessments are easy to conduct but are often incomplete. Inside-out assessments can be conducted either as a paperwork exercise like a questionnaire, or as a technical assessment by auditors, penetration testers, or red teams. The mapping for capabilities in this category align against the **IDENTIFY** function across all the assets within the vendor owned asset layer. The subfunction is vulnerability identification, rather than inventorying or classifying, because the goal of these assessments is to determine if vendors leave their assets in a vulnerable state, thereby introducing added risk to their customers.

Mapping Controls Testing

We also conduct various forms of testing to validate that we have the right controls and verify that these controls are set up correctly to secure our environment. The mapping of the testing activity depends on the security function being evaluated and which specific assets are subject to those controls. The following four controls testing activities are common across many organizations; however, they are often misunderstood with respect to their differentiation and intent. Mapping these activities to the Cyber Defense Matrix makes the finer distinction among these activities much clearer.

Vulnerability scanning/assessment. Vulnerability scanning is an **IDENTIFY** activity, testing our configuration controls. It answers the questions, "Did we configure or build everything properly?" and "Did we do it in a way that is free of vulnerabilities?" The asset class mapping depends on the asset being scanned or evaluated. For example, technologies that scan for misconfigured open file shares or publicly accessible S3 buckets would be mapped to **DATA-IDENTIFY**. The skill set to search for zero-day vulnerabilities in Windows would be mapped to **DEVICE-IDENTIFY**.

Penetration testing. Penetration testing is a **PROTECT** activity intended to assess our preventative controls. Presuming we have a vulnerability (patched or not), it answers the question, "Do we have controls in place which mitigate that vulnerability or prevent its exploitation?" Penetration testing is designed to determine if our environment can be penetrated by exploiting a vulnerability or if the vulnerability has been mitigated through other preventative controls. Penetration testing should not be asking the question, "Is there a vulnerability?" Apparently it often does, however, because a common complaint of those who purchase these services is that they have instead received a vulnerability scan. This is also the reason why it is wasteful to do a penetration test without having done a vulnerability scan first.[8]

The asset class that is being tested is important. We can break down each type of penetration testing to the underlying asset that is being evaluated. For example:

- Social engineering »→ **USER**
- Client side testing »→ **DEVICE**
- Web application testing »→ **APPLICATION**
- Wireless wardriving »→ **NETWORK**

Breach and attack simulation (BAS). Designed to test and assess detection controls, BAS falls into **DETECT**. The question it is answering is, "If I have a vulnerability and it is exploited, will the attack be successful?" But more importantly, "Will I know?" BAS is designed to test whether the alarms I set go off as they are supposed to when the attack happens. That will tell me whether an actual attack will be spotted. With BAS, many tests are designed to assess "inside out" controls, i.e., those alarms designed to prevent the exfiltration of data. Is the exfil activity itself logged as an event? Do I have filters that flag it as an alert? Does the alert get to the systems that a human analyst will look at? In other words, is it logged, flagged, and delivered? When an alert is delivered as part of a BAS, it is suppressed so as not to waste the analyst's time responding to a fictional intrusion.

Red team. These activities test responsive controls and fall under **RESPOND**. They answer the questions, "If an intrusion occurs, will my security team notice and act on it?" and "Will they be able to get rid of it?" Unlike in a BAS, I do not suppress an alert triggered by a red team intrusion. I want the analyst to see it so I can assess whether they identify it correctly and respond appropriately.

This is what a red team is for: to test response controls enacted by the blue team when an intrusion is discovered. I would argue that a red team should not need to have people skilled in vulnerability assessment or penetration

---

[8] Penetration testing can include an independent vulnerability assessment, but it does not require one. A penetration test should be informed by the results of previous vulnerability scans. If an independent vulnerability assessment is conducted and it reveals previously undiscovered vulnerabilities, this points to a need to improve vulnerability scanning capabilities.

testing to be successful. It is commonly assumed that red teams must have people who know how to discover new vulnerabilities and break into the perimeter of an organization. But, in my view, red teams simply need to trigger the alerts designed to drive blue team response actions.

Red teams should remember that their primary responsibility is to test response controls, not the other controls. I do think that red team members are better at what they do when they understand all four controls-testing functions, but they should rely upon existing tools and capabilities to test those functions and only supplement as needed when those tools and capabilities lack sufficient coverage. There should be no expectation that red teamers can find the latest zero-day.

CHAPTER 4

Security Measurements

> *When a measure becomes a target,*
> *it ceases to be a good measure.*
> – Goodhart's Law

Demonstrating Value When No News Is Good News

Security is a practice where we often feel the need to constantly prove our value in the absence of loss events. It is easy to demonstrate the value of a missing security control in hindsight after a breach. It is harder to convey that same value beforehand. Often, our best day is one where, to the rest of the organization outside of the security team, it seems like nothing happened. Because of this situation, we often turn to measurements to help us showcase our day-to-day value to stakeholders and to give ourselves assurance that our security controls are working as intended.

However, even with a wide array of measurements, it is hard to demonstrate this value or to understand where we have control deficiencies. Furthermore, we are constantly tempted to game the measurements to paint a rosier picture — to play up our successes and avoid confrontation. Using measurements in this way is risky, creating the danger of believing our own hype and giving ourselves a false sense of assurance about our security posture or operations. Our measurements may look good, but how is this truly useful if the measurements were cherry-picked and designed to make us look good in the first place?

Capturing meaningful security-related measurements is a common struggle. And as Goodhart suggests, even if we do find that ideal measurement, once we make it our goal, it no longer becomes ideal. There are plenty of great books on how to craft good measurements.[1] This book is not one of them. Instead, this chapter explains how to use the Cyber Defense Matrix in the following ways to avoid fixating on a limited set of measurements that may distort the big picture view of our security controls:

1. Organize measurements that we already have so that we can understand where we have controls and associated measurements.
2. Spot measurement gaps (and possible missing controls) and derive the type of measurements we may need.
3. Understand the depth/utility/quality of each available measurement in relation to one another.
4. View a broad set of measurement choices in context to avoid the pitfalls of poor decision-making due to narrow framing bias.
5. Consider the bad data as well as the good, and provide a comprehensive basis for telling a complete security story to ourselves and stakeholders.

Organizing Measurements

Because the Cyber Defense Matrix provides a comprehensive framework to look at a security environment, we can also use it to organize our measurements. Let us say we have a measurement that reflects how many endpoints in our organization are regularly scanned for vulnerabilities. That measurement fits in the **DEVICE-IDENTIFY** box. A measurement about the completion status of employees undergoing regular security refresher training fits in the **USER-PROTECT** box, and so on. In this way, we can organize the measurements we currently collect and align them against the security controls within our environment.

The aim of this sorting process is to have all our existing measurements in some box of the Cyber Defense Matrix. Along the way, we may discover that some of our existing measurements lack sufficient precision to be placed in a particular box. For example, let us suppose we have a mean time to detection (MTTD) measurement. This clearly falls somewhere in

---

[1] For example, Douglas Hubbard and Richard Seiersen's How to Measure Anything in Cybersecurity Risk; Jack Jones and Jack Freund's Measuring and Managing Information Risk; and Richard Seiersen's Metrics Manifesto.

the **DETECT** column, but which asset class should it align against? Well, that depends upon what type of compromise we are looking for. We should not expect the MTTD to be the same for an insider threat (**USER-DETECT**) and an endpoint compromise (**DEVICE-DETECT**). When we encounter measurements that are ambiguous in their mapping, we should consider refining them to add the specificity needed to map them into the Cyber Defense Matrix.

If we have used the Cyber Defense Matrix to map other parts of our security program, such as our technologies in use, organizing these measurements into the Cyber Defense Matrix allows us to see how these measurements can tie into these other components. For example, we should expect to see one or more of the technologies that are mapped to the **DEVICE-IDENTIFY** box provide the means to measure how many endpoints are regularly scanned for vulnerabilities. Processes and policies mapped to the **DEVICE-IDENTIFY** box should cover how often vulnerability scanning occurs. Skillsets found in the **DEVICE-IDENTIFY** box should support the ability to conduct or manage vulnerability scans. The Cyber Defense Matrix allows us to align functional needs across *PEOPLE*, *PROCESS*, and *TECHNOLOGY* to ensure that each part supports the whole.

Spotting Gaps

As available measurements are placed into the various boxes of the Cyber Defense Matrix, it will quickly become visually evident where gaps might exist. These are the boxes in the Cyber Defense Matrix that do not have any measurements in them. In our example above, we have a measurement on vulnerability scanning for **DEVICE-IDENTIFY**, but what about vulnerability scanning measurements for **NETWORK-IDENTIFY, APPLICATION-IDENTIFY, DATA-IDENTIFY**, and **USER-IDENTIFY**? Gaps like these might be due to a variety of factors, from a missing control to simple oversight.

The empty boxes we discover in the matrix can serve as a prompt to check if the underlying security technologies that also align to those empty boxes could be used as a source for measurements. In other words, if we happen to have a security technology that maps to a box where we have no measurements, we might want to investigate whether or not that technology can supply us with the measurements that we need.

But the matrix can also help us derive measurements for those empty boxes even if we do not actually use any security technologies which map to that box, or if those technologies do not provide any measurements themselves. Following the pattern matching framework provided by the matrix, we can look at measurements for other asset classes in the same column as our empty box to find hints for deriving a new measurement.

Let us take as an example **DATA-RESPOND** and imagine that we have no measurements or technologies available for that box. As shown in Figure 8, we can use the structure of the Cyber Defense Matrix to examine the measurements of other asset classes in the **RESPOND** column to find common themes that might help us define a measurement for **DATA-RESPOND**.

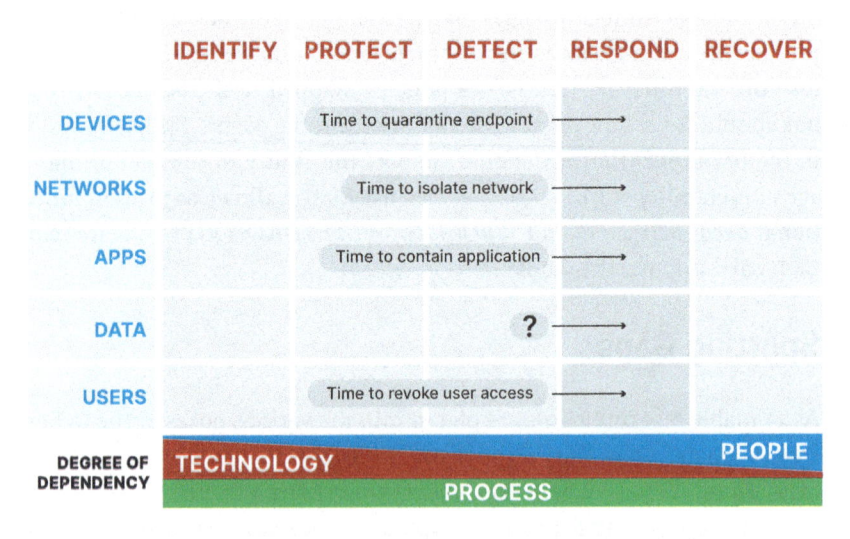

	IDENTIFY	PROTECT	DETECT	RESPOND	RECOVER
DEVICES			Time to quarantine endpoint →		
NETWORKS			Time to isolate network →		
APPS			Time to contain application →		
DATA			? →		
USERS			Time to revoke user access →		

DEGREE OF DEPENDENCY — TECHNOLOGY — PROCESS — PEOPLE

Figure 8: Using Pattern Matching to Discover Potential Measurements

For example, on the right of boom, time is a critical factor — something has blown up and we are trying to limit further damage. A measurement of our ability to quickly contain the damage blast radius would be expected across various asset classes. In the **DEVICE** asset class, we might have an ability to measure how quickly we can quarantine an infected endpoint. For **NETWORK-RESPOND**, it might be a measurement of how quickly we can isolate or firewall off that network. For **APPLICATIONS**, we might want to know how quickly we can contain or restrict attacker activity without having to shut down the exploited application. For **USERS**, we might have measurements for how quickly we can revoke the access of an employee or contractor who has become an insider threat.

Using pattern-matching, we can easily see that speed — how quickly we can isolate corrupted or compromised **DATA** so that it does not spread — is one measurement we could use for that **DATA-RESPOND** box. In this way, the organizational structure of the Cyber Defense Matrix can help us develop new measurements based on existing measurements found in other boxes of that column.

Measuring Measurement Quality

Not all measurements are created equal. There are classes of measurements that are easy to capture, but offer little value. Conversely, there are others that provide significantly more value, but are also much harder to obtain. In some cases, a measurement that is harder to obtain may have dependencies on measurements that are easier to obtain. As a rule, it makes the most sense to obtain lower-difficulty measurements before aiming higher. Figure 9 provides an example of how different types of measurements might be considered in relation to one another.

The five different levels of measurement offered in the diagram are an example of one way to calibrate the quality of the measurements used by an organization. We often struggle trying to obtain measurements that may be out of reach. Understanding what level of measurements we have available may reduce the frustrations we encounter as we seek to increase the strength of our measurement program.

I will offer an explanation of each level of measurement, with a vaccine analogy to make it more relatable. There are

HARD

⑤ **Efficiency**
How cost efficient is the capability? $$

④ **Performance**
How well does the capability work?   96%

③ **Utilization**
Are all available features enabled?

② **Coverage**
Is the capability enabled?

① **Presence**
Does the capability exist?

EASY

RELATIVE DIFFICULTY

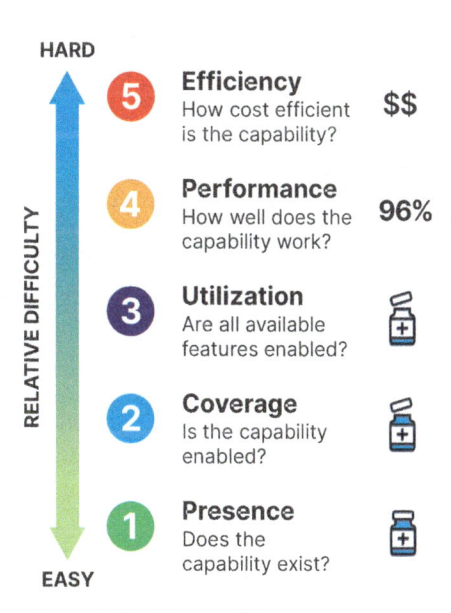

Figure 9: Five Classes of Measurements

likely other levels of measurement that are not represented, but overall, this type of construct can help us understand the different kinds of measurements we can capture that assess the strength of our security controls in any given box in the matrix.

Presence measurements. Presence measurements can be a binary indicator of whether or not a control is available. If it is a technology-centric control, it should reflect whether or not we have bought or built the technology to enforce that control. If it is a process-centric control, then it should reflect whether or not we have defined that process. Example measurements include:
- Do we have an endpoint protection platform (EPP)?
- Do we have a vulnerability management program?

Vaccine analogy: Is there a vaccine?

Coverage measurements. Coverage measurements are usually represented as a percentage of assets or targets which are subject to a given control. Unfortunately, the true denominator may not always be exactly known, but approximations are often sufficient when the denominator changes frequently. Coverage measurements should reflect the proportion of assets that are subject to the controls for a given box. Example measurements include:
- What percentage of endpoints have EPP installed?
- What percentage of our assets are covered under the vulnerability management program? Are there assets whose vulnerabilities are handled through a different process?

Vaccine analogy: How broadly has the vaccine been distributed? How many people have gotten at least one vaccine shot? How many have gotten vaccine shots from manufacturer X vs manufacturer Y?

Utilization measurements. Utilization measurements, usually represented as a percentage, capture the extent to which specific features and capabilities of a control are being put to use. Example measurements include:
- Are we using all the EPP features (e.g., antivirus, malware prevention, host firewall, host-based intrusion prevention system, etc.)? What percentage does that represent of the full set of available features?

- What percentage of reported vulnerabilities for a given asset class are handled by our vulnerability management program?

Vaccine analogy: What percentage of doses in each vial have been used? What percentage of people have gotten both shots?

Performance measurements. Performance measurements help us understand whether or not we are getting the security outcomes that we desire. Performance measurements are what many of us desire at the outset, but they are difficult and time-consuming to obtain.[2] If some of the previous measurements are not available, capturing performance measurements will be much more challenging. To accurately assess security outcomes, we need a foundation of understanding of the control's coverage and capabilities. To arrive at performance measurements, we also need to know if the controls are being properly operated and administered. A firewall covering all our assets does not provide much security benefit if the access control list is set to ANY-ANY (allowing everything). Example measurements include:

- How effective is this EPP capability in thwarting APT40-style attacks against my endpoints with as few false positives as possible?
- How much faster are we fixing our Windows vulnerabilities because of our vulnerability management program?

Vaccine analogy: How well does the vaccine work in preventing infections and reducing the severity of sickness when a booster shot is administered six months apart?

Efficiency measurements. Efficiency measurements help us understand if we are getting the desired outcomes in a cost efficient manner with the least waste and the best use of time and money. We all have limited resources, so efficiency measurements can help us understand trade-offs across the entire set of controls that we see through the Cyber Defense Matrix. Example measurements include:

- How efficiently can I reduce risk by investing in **DEVICE**-centric controls versus **NETWORK**-centric controls?
- What is the appropriate maximum time frame for fixing medium

---

2 Consider how long it takes for vaccines to be deemed efficacious and safe for use. And that is with a well established testing methodology!

severity vulnerabilities without creating an onerous burden on the rest of the organization?

Vaccine analogy: Can I reduce risk more efficiently with a one-shot vaccine with lower performance compared to a two-shot vaccine with higher performance but more overhead costs?

The Center for Internet Security (CIS) offers a starter set of measurements through their Controls Assessment Specification effort.[3] These measurements correspond to the CIS Critical Security Controls. Because the Critical Security Controls align with the Cyber Defense Matrix, these measurements can be directly mapped to the matrix as shown in Figure 10. Each large number represents the total number of measurements that have been cataloged in the Controls Assessment Specification effort. The smaller numbers represent measurements associated with the three implementation groups defined by CIS.[4] The first of the three numbers are measurements that anyone should be able to collect. The second and third numbers reflect implementation groups 2 and 3 respectively, which correspond to organizations that have progressively more resources to address security issues.

	IDENTIFY	PROTECT	DETECT	RESPOND	RECOVER	TOTAL
DEVICES	15 (8/7/0)	84 (30/36/18)	26 (0/21/5)	9 (5/4/0)	0	134 (43/68/23)
NETWORKS	14 (0/10/4)	94 (21/53/20)	32 (3/28/1)	0	0	140 (24/91/25)
APPS	56 (18/34/4)	125 (33/72/20)	5 (0/5/0)	11 (7/4/0)	0	197 (58/115/24)
DATA	22 (9/13/0)	67 (32/21/14)	12 (0/0/12)	0	17 (13/4/0)	118 (54/38/26)
USERS	24 (9/15/0)	90 (84/6/0)	0	6 (6/0/0)	0	120 (99/21/0)
TOTAL	131 (44/79/8)	460 (200/188/72)	75 (3/54/18)	26 (18/8/0)	17 (13/4/0)	709 (278/333/98)

Figure 10: Number of Measurements from CIS' Controls Assessment Specification

---

3 https://controls-assessment-specification.readthedocs.io/en/stable/
4 https://www.cisecurity.org/controls/implementation-groups/

There are a total of 709 measurements identified by the CIS Controls Assessment Specification effort, each generally corresponding to one of the five levels of measurement described in Figure 9. As we collect our own measurements, characterizing the measurement level will allow us to see our overall progression of measurement quality, as shown notionally in Figure 11.

	IDENTIFY	PROTECT	DETECT	RESPOND	RECOVER
DEVICES	E	E	D	E	F
NETWORKS	C	D	B	F	F
APPS	E	E	A	B	F
DATA	E	D	B	F	E
USERS	B	C	F	E	F

Figure 11: Characterizing a Security Program's Overall Measurement Quality

This shows at a glance how robust our overall measurements might be. Boxes with an "F" can reflect where we lack any measurements while boxes with an "A" or "B" will convey that the measurements will likely to be more useful for making decisions about the controls associated with those boxes.. If there are multiple measurements within a box, we could calculate an average or a range.[5]

Seeing the Big Picture

A common cognitive bias that can lead to poor outcomes is narrow framing, which is when decisions are made based on the details of a specific circumstance without the context of the broader choices that are available. When we look at measurements in isolation and seek to improve a particular measurement without consideration of the bigger picture, we fall into the trap of narrow framing. This happens frequently when an

---

5 This may be better depicted as a box plot diagram, but it could get very complicated-looking very fast.

activity that we are tracking (e.g., missed SLA for patching) turns red on a dashboard and we become laser focused on turning that activity back to green without considering the broader context of why it turned red to begin with.

Because the Cyber Defense Matrix forces us to look at measurements from a broader perspective, it helps avoid narrow framing and ensures that we are seeing the bigger picture. We can directly compare the strength of our program activities from one box to another and make smarter decisions about where to invest resources, focusing on those areas where we are likely to get the highest marginal return.

It also exposes any counterfactual data — measurements that might challenge our narratives about how strong our security program is and reveal any weaknesses it might have. By helping us organize a comprehensive corpus of measurements across the whole security environment, the Cyber Defense Matrix can ensure we avoid the trap of believing our own hype.

Let us consider hygiene, for instance. The measurements we choose to show might reflect that our hygiene is great on our endpoints — our **DEVICES**. But what about our hygiene as it relates to the other four asset classes? How is the hygiene on our **APPLICATIONS, NETWORKS, DATA**, and **USERS**? By forcing us to look at measurements from across the whole security environment, the matrix ensures that we look at the bad as well as the good.

This forcing function is also useful when making the case for security funding. We often need to make the case for more resources, but we sometimes find ourselves trying to walk a tightrope when it comes to measurements. We want our data to show that we are doing a great job in securing the organization while simultaneously having the data show gaps or holes in our security posture to support our need for additional funding.

The Cyber Defense Matrix helps resolve this dilemma by providing a structure we can use to explain to stakeholders that we're doing well in some boxes while still needing to invest further in others. It provides a ready-made structure for a narrative that tells both sides of the story, showing stakeholders where security is strong and where it needs additional investment.

Keeping Measurements in Perspective

By providing a comprehensive framework for measurements across our whole environment, the Cyber Defense Matrix helps ensure that we have all the data we need in a well-organized fashion for any measurements we want to produce. At the same time, it is important to approach measurements carefully. We security practitioners are fond of touting measurements to show progress towards security goals, even ambiguous ones. However, we need to remind ourselves of Goodhart's law and avoid making a measurement a target.

When it comes to measuring security, we come across more data than we probably realize. The challenge is that we often do not know what to do with that data or how it connects to a bigger picture. The Cyber Defense Matrix is designed to help us discover and organize consistent, repeatable measurements that over time will provide the data we need to build an accurate picture of how our security program is progressing.

CHAPTER 5

Developing a Security Roadmap Using the Stack

*If you do not know where
you're going, any road
will take you there.*

— Lewis Carroll

Playing Security Bingo

If you have ever been to RSA, Black Hat, or one of the other big cybersecurity conferences, you are familiar with the game of security bingo. Players wander around the exhibit hall with a card bearing a grid of the trendiest marketing terms that year. Every time you hear one of those terms from a salesperson on the exhibit floor, you can cross it off your card. The winner is the first one to get every box on the grid crossed off.

In this chapter, we will use the Cyber Defense Matrix to play security bingo for real. And the goal here is not necessarily to play blackout — we do not have to get every box filled in to win. Indeed, most enterprises cannot afford to fill every box on the matrix with security technologies. They may not have the budget, staff, or resources to do that. Instead, the goal in this game of bingo, and the aim of this chapter, is to help us make better risk management decisions about which boxes we really need to cover, and which we can afford to leave uncovered or lightly covered. Every organization's risk tolerance is different, but the structure of the Cyber Defense Matrix enables us to break down those risk assessments for each box in the matrix. Our goal is to develop a roadmap that gives us a security posture that is good enough to be able to declare bingo.

When we seek to develop a security roadmap, we generally want to answer three questions:

1. How secure are we now?
2. How secure should we be?
3. How do we get from here to there?

Answering these questions requires us to consider multiple factors, such as our current state, the requirements that we have to meet, the risk environment, available capabilities, and business constraints. As shown in Figure 12, each of these factors correspond to a layer in a construct that I call the "Stack."

Foundation
Cyber Defense Matrix

Layer 1: Pantry
Current State Capabilities

Layer 4: Market
Commercial Options, Art of the Possible

Layer 2: Recipes
Proven Practices, Requirements, Compliance Frameworks

Layer 5: Allergies
Business and Technology Constraints, Exceptions

Layer 3: Nutritional Needs
Risk Environment, Attack Surfaces, Threat Landscape

The Stack
Combined Matrices

Figure 12: The Five Layers of the "Stack"

Each layer uses the Cyber Defense Matrix as the foundation to organize information associated with that layer. It can be difficult to grasp how all these factors relate to each other, so I will return to the food analogy we used in Chapter 1. Using a meal preparation analogy may help you remember all the different steps and understand the relationships among them, but the intention is not to tie your hands to doing things in a particular order. Just as in real life, you might visit the store without a recipe in mind, so when using the matrix to build your version of the Stack, you do not need to slavishly follow the order of these steps. They are designed to provide flexibility, not constraints.

In fact, you do not even have to complete all of the steps outlined here. You can get helpful insights for your roadmap based on populating only a few of these layers. However, I encourage you to at least attempt them all. If you are intentional about using the Cyber Defense Matrix as a mental model to organize information as you encounter it, over a short period of time, you will find all the layers of your Stack to be well populated.

Layer 1: Checking the Pantry (Current State)

For this first layer, we map our existing portfolio of security capabilities to the matrix to produce something like Figure 13. Think of this as checking to see what you have in the pantry or cupboard. Chapter 3 provides guidance for doing this, so you may want to review it if you are not sure how it works. If the organization consists of many subsidiaries or departments, each sub-organization can map its capabilities to the matrix, and we can combine these individual mappings to determine where we have common technological capabilities across the organization.

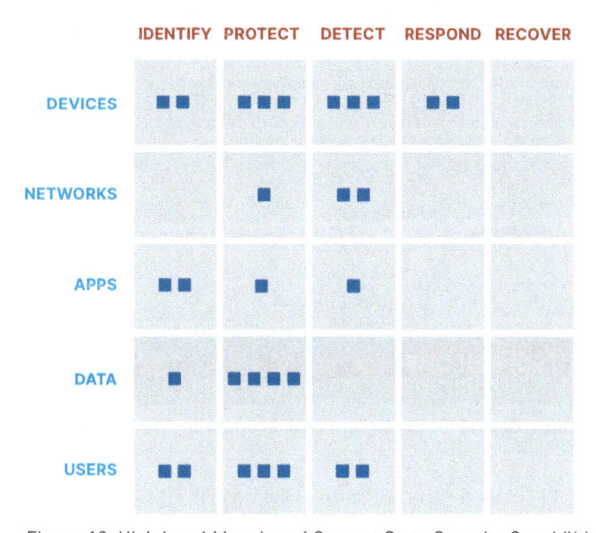

Figure 13: High Level Mapping of Current State Security Capabilities

Some gaps may already be evident according to the Cyber Defense Matrix structure, but that does not necessarily mean that there is a problem. Conversely, we may find many capabilities within a given box of the matrix — meaning that we may have too many similar capabilities in that area (having too much of one type of ingredient in the pantry). If you find yourself in this situation, I will discuss how we can rationalize our technological capabilities in Chapter 8.

Layer 2: Finding Recipes (Requirements)

A recipe is a set of requirements for a meal — a list of all the things we will need to prepare it. In developing a security roadmap, we need to understand a similar set of requirements for our security program. These are proven practices as defined through reference architectures and compliance frameworks. Figure 14 provides an example of such an alignment showing the Center for Internet Security's (CIS) Critical Security Controls aligned against the Cyber Defense Matrix. This enables us to more easily identify where our most important security needs are, and it enables pattern matching comparisons across the five functions and five asset classes.

	IDENTIFY	PROTECT	DETECT	RESPOND	RECOVER
DEVICES	1.1, 1.4	3.6, 4.4, 4.5, 4.8, 4.9, 4.11, 4.12, 10.1, 10.2, 10.3, 10.5, 10.6, 12.7, 12.8, 13.5, 13.7, 13.9	1.3, 1.5, 8.8, 10.4, 10.7, 13.2	1.2, 4.10	
NETWORKS	12.4, 18.1, 18.2, 18.5	3.12, 4.2, 4.6, 8.1, 8.3, 8.4, 8.10, 9.2, 9.3, 9.5, 9.6, 9.7, 12.1, 12.2, 12.3, 12.5, 12.6, 13.4, 13.8, 13.10, 18.3, 18.4	8.2, 8.5, 8.6, 8.7, 8.9, 8.11, 13.1, 13.3, 13.6, 13.11		
APPS	2.1, 2.2, 7.5, 7.6, 15.1, 15.2, 15.3, 15.5, 18.6, 18.7, 18.8	2.5, 2.6, 2.7, 4.1, 7.1, 7.3, 7.4, 9.1, 9.4, 15.4, 16.1, 16.2, 16.3, 16.4, 16.5, 16.6, 16.7, 16.8, 16.9, 16.10, 16.11, 16.12, 16.13, 16.14, 18.9, 18.10	2.4	2.3, 7.2, 7.7	
DATA	3.1, 3.2, 3.7, 3.8	3.3, 3.4, 3.5, 3.9, 3.10, 3.11, 3.13, 6.8, 11.3, 14.6, 15.7, 18.11	3.14, 8.12, 15.6		11.1, 11.2, 11.4, 11.5
USERS	5.1, 5.5, 6.6	4.3, 4.7, 5.2, 5.4, 5.6, 6.1, 6.2, 6.3, 6.4, 6.5, 6.7, 14.1, 14.2, 14.3, 14.4, 14.5, 14.7, 14.8, 14.9		5.3	

DEGREE OF DEPENDENCY: TECHNOLOGY → PEOPLE / PROCESS — 17.1, 17.2, 17.3, 17.4, 17.5, 17.6, 17.9, 17.7, 17.8

■ Implementation Group 1   ■ Implementation Group 2   ▨ Implementation Group 3

Figure 14: CIS' Critical Security Controls Mapped to the Cyber Defense Matrix[1]

An organization may be subject to numerous requirements from multiple compliance frameworks. Organizing requirements into the Cyber Defense Matrix will make it easier to see how different frameworks map

---

[1] https://www.cisecurity.org/controls

to each other and where there may be overlapping or even conflicting requirements.

Moreover, if we have done the first layer, we can now overlay these two layers to see where we might have actual gaps from a compliance perspective as shown in Figure 15. By overlaying our existing portfolio mapped in the previous step with the compliance requirements in this layer, we can easily see where we have potential gaps — areas where our security is deficient or not being addressed at all by our existing security capabilities. This is like discovering that ingredients are present in inadequate quantities or missing altogether from the pantry.

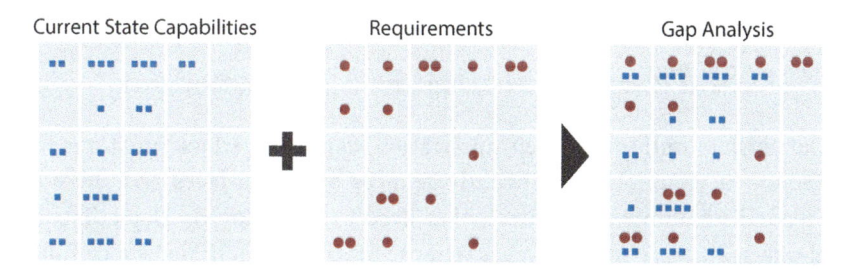

Figure 15: Gap Analysis Through a Comparison of Current State with Requirements

Layer 3: Defining Our Nutritional Needs (Risks)

The next step is to map our risks. Think of this step as defining our daily nutritional needs based on what we care about (high value assets), our lifestyle (attack surfaces), and the current best scientific understanding of what constitutes a healthy diet (threat intelligence about current attacker behavior.) Mapping our vulnerable attack surfaces, where we have high value assets, and our current threat environment to the matrix helps to highlight those areas that deserve additional attention because attackers may be focused on them.

There are a few possible ways to conduct this mapping. For example, attack surfaces are determined based on the choices that we make in how we build our infrastructure. A traditional three-tiered web application might have the attack surfaces as shown in Figure 16. Because of these attack surfaces, we would need to have controls to address each of these points of exposure.

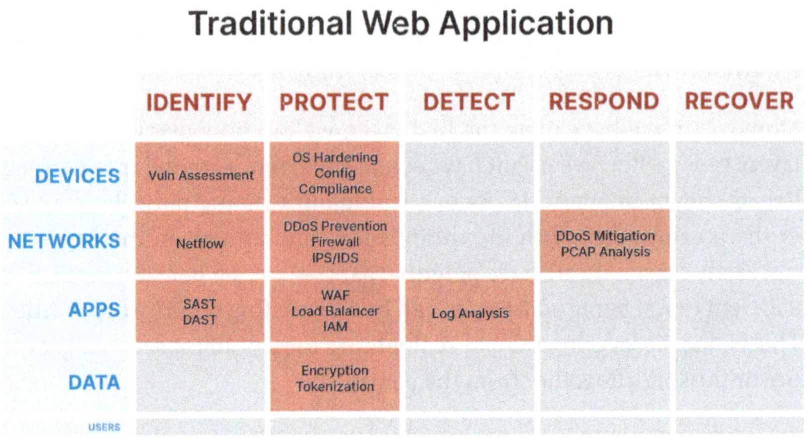

Traditional Web Application

	IDENTIFY	PROTECT	DETECT	RESPOND	RECOVER
DEVICES	Vuln Assessment	OS Hardening Config Compliance			
NETWORKS	Netflow	DDoS Prevention Firewall IPS/IDS		DDoS Mitigation PCAP Analysis	
APPS	SAST DAST	WAF Load Balancer IAM	Log Analysis		
DATA		Encryption Tokenization			
USERS					

Figure 16: Attack Surfaces of a Traditional Three-Tiered Web Application

Conversely, building a web application using serverless functions, as shown in Figure 17, significantly reduces attack surfaces and the corresponding number of controls that need to be deployed.[2] Depictions of attack surfaces using the Cyber Defense Matrix can help make a compelling case for adopting new design patterns that build on more defensible infrastructure.

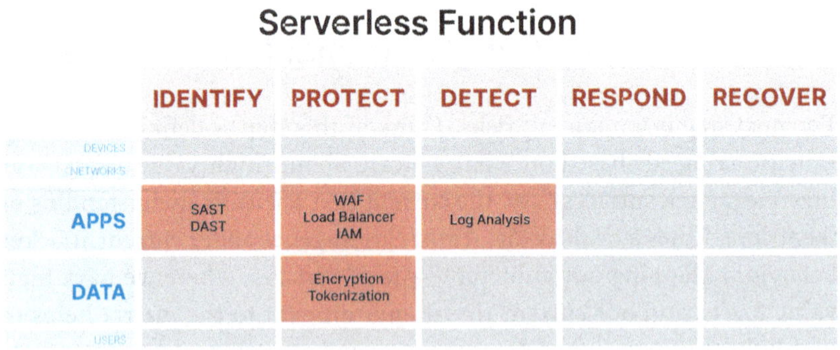

Serverless Function

	IDENTIFY	PROTECT	DETECT	RESPOND	RECOVER
DEVICES					
NETWORKS					
APPS	SAST DAST	WAF Load Balancer IAM	Log Analysis		
DATA		Encryption Tokenization			
USERS					

Figure 17: Attack Surfaces of a Serverless Function

Another method of mapping risks to the Cyber Defense Matrix is to capture the threat environment. Constructs like CIS' Community Defense

---

[2] These attack surfaces do not technically go away, but are transferred to whomever is operating the underlying serverless function infrastructure.

Model[3] or Ross Young's Threat and Safeguard Matrix[4] can be used to combine specific threats, such as ransomware, web application attacks, and phishing to the controls and safeguards that are best able to thwart these attacks. Figure 18 provides an example of the CIS controls that are best suited to address ransomware.

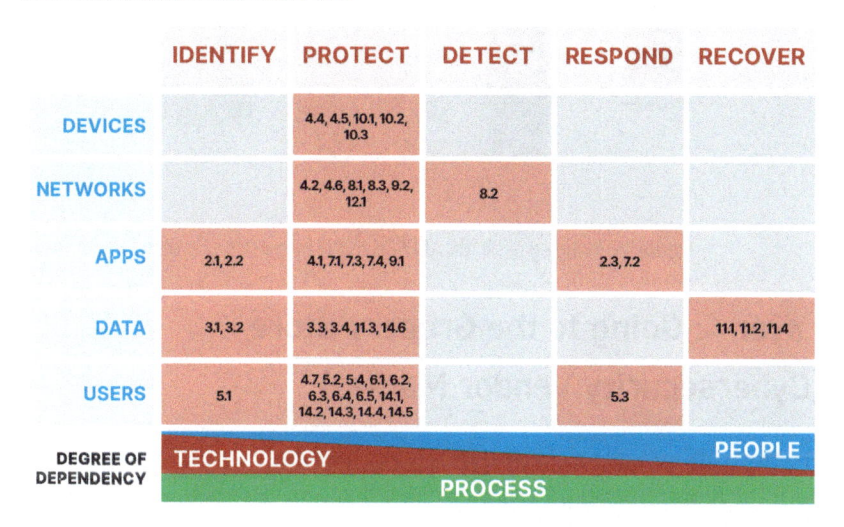

	IDENTIFY	PROTECT	DETECT	RESPOND	RECOVER
DEVICES		4.4, 4.5, 10.1, 10.2, 10.3			
NETWORKS		4.2, 4.6, 8.1, 8.3, 9.2, 12.1	8.2		
APPS	2.1, 2.2	4.1, 7.1, 7.3, 7.4, 9.1		2.3, 7.2	
DATA	3.1, 3.2	3.3, 3.4, 11.3, 14.6			11.1, 11.2, 11.4
USERS	5.1	4.7, 5.2, 5.4, 6.1, 6.2, 6.3, 6.4, 6.5, 14.1, 14.2, 14.3, 14.4, 14.5		5.3	

DEGREE OF DEPENDENCY — TECHNOLOGY ... PEOPLE / PROCESS

Figure 18: Mapping of Ransomware to CIS Controls and the Cyber Defense Matrix

The threat intelligence offered by most vendors map to the MITRE ATT&CK Framework.[5] As such, a cross-reference between MITRE ATT&CK and the asset class that is being targeted can help in placing attacker observations into the Cyber Defense Matrix. Figure 19 shows a notional mapping, aligning the various tactics, techniques, and procedures (TTP) in MITRE ATT&CK with the associated asset class. The MITRE ATT&CK Framework points out various log sources that can be consumed to support the function of **DETECT** and mitigations that can counter the attacker TTPs to perform the function of **PROTECT**. These references can subsequently be used to map specific TTPs to the Cyber Defense Matrix.[6]

---

3 https://www.cisecurity.org/insights/white-papers/cis-community-defense-model-2-0
4 https://owasp.org/www-project-threat-and-safeguard-matrix/
5 https://attack.mitre.org.
6 There is a danger in taking a strategic model, such as the Cyber Defense Matrix and trying to map it to a more tactical model, like the MITRE ATT&CK Framework. Specifically, it will be easy to lose the proverbial forest for the trees when the Cyber Defense Matrix is used to map tactical level information.

Figure 19: Mapping of MITRE ATT&CK to the Five Asset Classes[7]

Layer 4: Going to the Grocery Store (Cybersecurity Vendor Market)

Having captured our current state capabilities and comparing them with both the requirements and risks, we now have a better perspective of gaps that need to be addressed. We can think of this as shopping for items listed on the recipe that are not already on hand in the pantry. Where there are gaps or areas of insufficient coverage left by our existing security capabilities, we can look at the full range of options available in the security marketplace to fill them.

Gaps that cannot be addressed by commercial vendors can feed into R&D requirements. Alternatively, if our organization has sufficient resources or a compelling need, we may try to build our own internal capability to provide coverage in that area. Figure 20 shows a sample of various security product categories mapped to the Cyber Defense Matrix. (The stars are meant to represent the breadth of choices within a given box, not the quality of the choices.)

---

7 This is an older version of the MITRE ATT&CK Framework, but you get the idea.

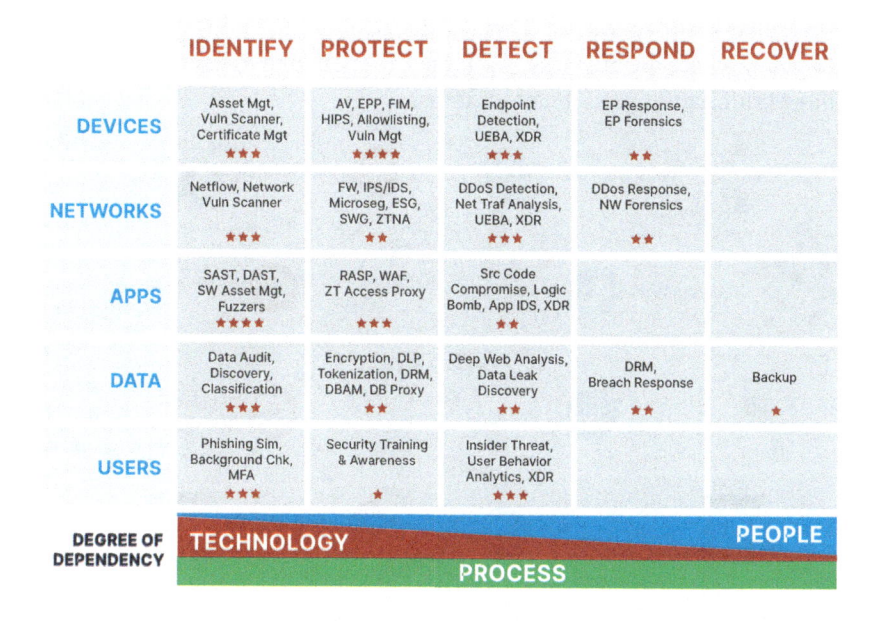

	IDENTIFY	PROTECT	DETECT	RESPOND	RECOVER
DEVICES	Asset Mgt, Vuln Scanner, Certificate Mgt ★★★	AV, EPP, FIM, HIPS, Allowlisting, Vuln Mgt ★★★★	Endpoint Detection, UEBA, XDR ★★★	EP Response, EP Forensics ★★	
NETWORKS	Netflow, Network Vuln Scanner ★★★	FW, IPS/IDS, Microseg, ESG, SWG, ZTNA ★★	DDoS Detection, Net Traf Analysis, UEBA, XDR ★★★	DDos Response, NW Forensics ★★	
APPS	SAST, DAST, SW Asset Mgt, Fuzzers ★★★★	RASP, WAF, ZT Access Proxy ★★★	Src Code Compromise, Logic Bomb, App IDS, XDR ★★		
DATA	Data Audit, Discovery, Classification ★★★	Encryption, DLP, Tokenization, DRM, DBAM, DB Proxy ★★	Deep Web Analysis, Data Leak Discovery ★★	DRM, Breach Response ★★	Backup ★
USERS	Phishing Sim, Background Chk, MFA ★★★	Security Training & Awareness ★	Insider Threat, User Behavior Analytics, XDR ★★★		
DEGREE OF DEPENDENCY	TECHNOLOGY		PROCESS		PEOPLE

Figure 20: High Level Mapping of Security Categories and Availability of Commercial Solutions

Layer 5: Avoiding Allergens (Business Constraints)

The fifth layer represents allergies. This layer captures the factors that limit an organization's ability to implement security controls. These constraints can be business-oriented or technology-oriented, and they typically arise due to the criticality of specific functions (e.g., no downtime) or shifts in technology or architectures (e.g., end-to-end encryption, SaaS) that reduce the effectiveness of existing capabilities (e.g., deep packet inspection).

You can think of this as cataloging any allergens or other dietary restrictions that limit which ingredients you can include in your meal. As with food, there are legitimate business-driven constraints that prevent or limit the implementation of security controls in certain boxes of the matrix.

As shown in the following figure, these constraints can be mapped to the Cyber Defense Matrix to provide a broad perspective of where you might encounter resistance from the business when trying to implement a particular security control. This resistance often manifests as exceptions to policy where the business asks for permission to remain non-compliant due to some anticipated business impact.

Figure 21: High Level Mapping of Notional Business Constraints

Different business units may have different constraints. The Cyber Defense Matrix enables you to catalog these constraints and understand where it might affect your control implementation. You can also depict where constraints are negotiable (e.g., shaded boxes) or non-negotiable (e.g., black boxes). Different business drivers may create different constraints on your security control implementation. For example, you could easily implement employee monitoring controls in a call center but would have to negotiate implementing this same control for office workers, and it may be a non-starter for a senior executive.

One example of a non-negotiable business constraint might be removing local administrative rights for software developers. Locking down admin privileges is a best practice for **DEVICE-PROTECT**. But if we remove a developer's admin privileges, they often run into issues where they cannot deploy code, negatively impacting their productivity. Granting an exception to allow developers to retain local admin privileges is often a non-negotiable business constraint.

Another example of a non-negotiable business constraint would be creating latency in a high-speed commodity trading operation. The whole business is based on speed, so if we implement a security control that creates latency, we break the business model. It really does not matter how important that control is to our security — if we implement it, we will have a secure enterprise that does not work. It will not be able to fulfill the functions that make it a viable business. That means we might not be able to implement any controls on **NETWORK-PROTECT**. If slowing down the machines the traders use is also a no-no, we might face non-negotiable constraints on **DEVICE-PROTECT** as well. Each of these constraints would then be reflected on the Cyber Defense Matrix for that business unit.

The Stack

When you have two or more of these layers captured and mapped to the Cyber Defense Matrix, we can then organize these seemingly disparate pieces of information to be able to answer the three questions that we often seek from a roadmap, as shown in Figure 22.

Figure 22: Comparing All Five Layers to Get a Strategic View of the Problem Space

The Cyber Defense Matrix is intended to give a broad, strategic view of our problem space and this approach allows us to see it across multiple dimensions. It enables us to examine our security environment from all these individual layers. More importantly, it allows us to organize the information from these layers in a consistent manner so that we can then overlay each of these layers to arrive at a composite view as shown in Figure 23. This layered view is the Stack.

- Requirements
- Existing Capabilities
- Commercial Capabilities
- Risky Environment
- Negotiable Business Constraints
- Non-Negotiable Business Constraints

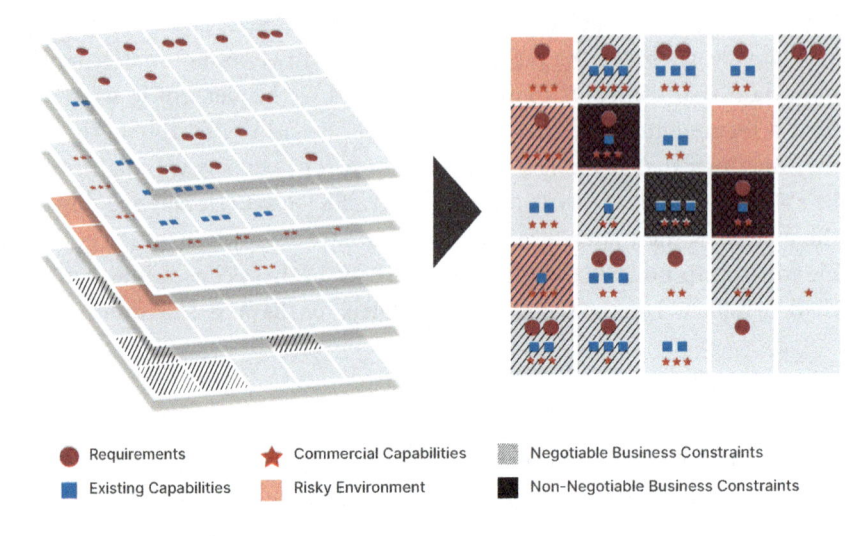

Figure 23: Combining the Five Layers to Create the Stack

Navigating the Stack

With the Stack, we can then plot our roadmap. If the symbology is unfamiliar, do not be distressed. Anyone who is unfamiliar with the symbology and legend of an aeronautical or hydrographic chart will be unable to properly navigate with it until they learn how to interpret the map. Similarly, it may take some time to properly interpret the Stack, but you should feel free to use your own symbology for the various layers that enable you to better understand the big picture.[8]

Given the notional example in Figure 24, we can use the symbology to quickly find potential paths to navigate through the cybersecurity landscape. Starting from the left, we want to progress through the five functions of the NIST Cybersecurity Framework.

---

8 Eventually, if this approach becomes popular, then we will want to normalize our symbology, but it would be too early to do that now.

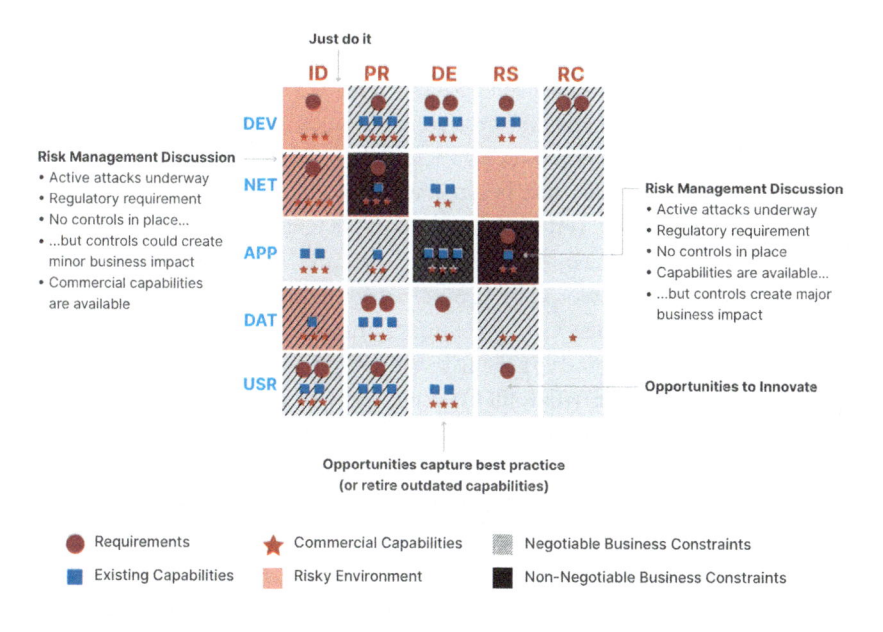

Figure 24: Interpreting Possible Actions from the Stack

First in the **IDENTIFY** column, we see across the five asset classes that there are constraints for three of them **(NETWORKS, DATA, USERS)**. Think of these as mountains in your cybersecurity landscape. We also see three areas that we have active, unaddressed risk in our environment **(DEVICES, NETWORKS, DATA)**. Think of these as waypoints that you want to make sure you visit. There are also compliance requirements that may force you to visit other areas as well **(DEVICES, NETWORKS, USERS)**. Fortunately, you have already visited a few of those areas and others, represented as existing capabilities **(APPLICATIONS, DATA, USERS)**. With this groundwork, we can understand what paths we have in front of us.

- In the **DEVICE-IDENTIFY** box, we have the combination of a compliance requirement and some unaddressed risk. Furthermore, there are several commercial capabilities available with no business interruptions if these capabilities are implemented. This is a "just do it" situation and offers a low hanging fruit opportunity to reduce risk without business impact.

- In the **NETWORK-IDENTIFY** box, we have the same situation as above, but this time with a potential minor business impact from implementing a control here. This will require a risk management discussion to

DEVELOPING A SECURITY ROADMAP   **69**

see if this is a mountain worth climbing. Depending upon the severity of the threat, it may be worth fighting to climb that mountain.

- In the **APPLICATION-IDENTIFY** box, we have an existing capability, but no requirement to have such a capability in place. There may be no unaddressed risk here because we have this capability, but that should be verified. If there is an acceptable level of risk here, this scenario may offer the opportunity to deprecate an unneeded control. Strategically, this would be equivalent to making the decision to take resources from one city to go conquer another city. However, consider also an alternative scenario where the business was unwilling to budge on the **NETWORK-IDENTIFY** box above and we were unable to implement a control there. These assets are all connected, so even though there may not be a requirement or risk here, it may make sense to fortify the **APPLICATION-IDENTIFY** box if it provides indirect support to the **NETWORK-IDENTIFY** box.

- In the **DATA-IDENTIFY** box, we have an existing capability, but no requirement to have such a capability in place. There is unaddressed risk here despite having some capabilities, and these capabilities create some level of business impact. The fact that there is already a capability here means that the business was willing to accept the business impact. We should check to see if the existing capabilities are being fully utilized and, if not, enable them if they counter the unaddressed risk. Otherwise, we may need to acquire additional capabilities from the commercial market.

- In the **USER-IDENTIFY** box, we have the combination of multiple requirements, existing capabilities to address those requirements, and no remaining unaddressed risk. These capabilities introduce business impact, but they have been accepted by the business. New initiatives in this box can take a lower priority, but if there are commercial capabilities that can eliminate the business impact, it will likely be welcomed by the business.

This is just a notional example of how we can interpret the Stack to help develop a roadmap tailored for our own organization to improve our security posture. We tend to think of our security environments as being unique—thus requiring a unique security stack. But in actuality, when you view a multitude of different environments through the consistent lens of the Cyber Defense Matrix, common patterns are quickly discernible, many driven simply by the nature of the business constraints that we all face. If we can understand these common patterns, we can share them with others who encounter similar business constraints and use them to determine where we might need new tools or where we might have under-rotated or over-rotated on our security controls. Hopefully, this will allow more of us to declare bingo for our security program.

CHAPTER 6

Improving Situational Awareness

*Where is all the knowledge
we lost with information?*

— T. S. Eliot

Seeing Threats in Context

Cyber situational awareness helps us to mitigate the loss or compromise of our assets. With higher levels of situational awareness, we can understand the state or behavior of our assets, and know how threats might be evading our security controls or exploiting vulnerabilities within our environment. But we are often challenged in our ability to consistently attain the needed levels of awareness. This chapter provides a framework-based approach for methodically and systematically improving our situational awareness to detect and respond to security incidents.

Understanding Situational Awareness

The most accepted definition for situational awareness comes from Mica Endsley's classic paper on Situation Awareness Theory.[1] She defines it as "the perception of the elements in the environment within a volume of time and space, the comprehension of their meaning, and the projection of their status in the near future." Based on this definition, we encounter several challenges to attaining higher levels of situational awareness.

---

1 Endsley, M. R. (1995). Toward a theory of situation awareness in dynamic systems. Human Factors, 37(1), 32-64.

Faulty visibility. To perceive something, we have to be able to see it. Oftentimes, we do not have the visibility that we need.

Faulty perception. Just because we see something does not mean that we are consciously aware of it. We have to perceive and notice that it is there. Information overload is a common cause of faulty perception.

Faulty comprehension. Even if we can see it and perceive it, we might not comprehend what we are looking at. We need to link and weave together the essential elements of the puzzle to complete the picture and project what may happen unless we take action.

A simple example will help illustrate the concepts of visibility, perception, comprehension, and projection. Figure 25 is an output of a forward web proxy that captures outbound web traffic. We get visibility from the log itself.

```
1   Fri 20 Nov 2015 14:37:07 PST: 134.173.42.70 http://sync.mathtag.c
2   om/sync/img?mt_exid=10025&redir=http%3A%2F%2Fsu.addthis.com%2Fred
3   %2Fusync%3Fpid&mm_bnc&mm_bct&UUID=223d-33d212 442   Sun 22 Nov 201
4   5 22:51:24 PST: 134.173.197.65 http://download.m0zilla.com/?produ
5   ct=firefox-42.0-complete&os=osx&lang=en-US 401   Sun 22 Nov 2015 2
6   2:51:25 PST: 134.13.197.6 http://download.cdn.m0zilla.com/pub/fir
7   efox/releases/42.0/update/mac/en-US/firefox-42.0.complete.exe 300
8   480   Sun 22 Nov 2015 22:57:59 PST: 134.173.197.65 http://www.find
9   evil.com/ 1888   Sun 22 Nov 2015 23:05:58 PST: 134.173.197.65 http
10  ://cs.hmc.edu/ 179   Tue 24 Nov 2015 10:07:05 PST: 134.173.42.70 h
11  ttp://self-repair.mozilla.org/en-US/repair 572   Tue 24 Nov 2015 1
12  0:07:25 PST: 134.173.42.70 http://www.pomona.edu/sites/default/fi
13  les/css/css_QsWyDNAFYyPOLo_fQ5W5McjIhuOqPPgAPPkIi9BpgrI.css 13296
14  Tue 24 Nov 2015 10:12:11 PST: 134.173.42.70 http://www.googletagm
15  anager.com/gtm.js?id=GTM-PVBCHG 17495   Tue 24 Nov 2015 10:12:11 P
```

Figure 25: Visibility - Web Proxy Log

But our visibility might be faulty if our logs are truncated, as shown in Figure 26; if we have incomplete visibility by only having a subset of our outbound Internet traffic going through a forward web proxy; or if logging is not enabled.

```
16  ST: 134.173.42.70 http://www.pomona.edu/sites/default/style
17  s/homepage_spotlight/public/spotlight.jpg?itok=-VKvyhfY 28169  Tu
18  e 24 Nov 2015 10:12:11 PST: 134.173.42.70 http://evil.com/ 654  T
19  ue 24 Nov 2015 10:12:11 PST: 134.173.42.70 http://www.google-anal
20  ytics.com/analytics.js 228  Tue 24 Nov 2015 10:12:11 PST: 134.173
```

Figure 26: Truncated Visibility - Web Proxy Log

These logs will not really mean anything until we can perceive their important elements. Figure 27 provides examples of what these might be. These elements might be discovered through pattern matching rules or filters, and these rules and filters will require constant tuning, ideally by those who have to deal with the corresponding output of those rules.

```
9   evil.com/ 1888  Sun 22 Nov 2015 23:05:58 PST: 134.173.197.65 http
```
```
18  e 24 Nov 2015 10:12:11 PST: 134.173.42.70 http://evil.com/ 654  T
```

Figure 27: Perception - Finding Evil

The tuning of these rules is critical. If this is not done well, we might miss some key bits of important information, such as mozilla being spelled with a zero instead of an O, and one of the supposedly evil sites perhaps not being evil after all, as shown in Figure 28.

```
6   2:51:25 PST: 134.13.197.6 http://download.cdn.m0zilla.com/pub/fir
7   efox/releases/42.0/update/mac/en-US/firefox-42.0.complete.exe 300
8   480  Sun 22 Nov 2015 22:57:59 PST: 134.173.197.65 http://www.find
9   evil.com/ 1888  Sun 22 Nov 2015 23:05:58 PST: 134.173.197.65 http
```

Figure 28: Faulty Perception

Next, we need to take what we can perceive and enrich it with other information so that we can comprehend what is going on and render a verdict. Threat intelligence provides visibility of threat actor assets and can be used to match against website domains in the proxy logs to discover potentially malicious domains (e.g., evil.com) that may have been visited by employees. If we have faulty visibility or faulty perception due to poorly configured filters and rules, we may block more than we intended (e.g., findevil.com) or completely miss suspect sites (e.g., m0zilla.com).

Once we have comprehension, the next level of situational awareness is being able to project what may happen if we do not take any action (e.g., lateral movement). This situational awareness informs what course of action we should take (e.g., block malicious/suspect domains and investigate endpoints that visited those domains).

Seeing Through Blind Spots with Frameworks

Faulty visibility, perception, or comprehension can hinder us from attaining situational awareness. However, sometimes we have blind spots that we do not even know we have. Frameworks address this problem by providing structure, helping us understand what constitutes completeness and track our progress towards it.

Using frameworks, such as the Cyber Defense Matrix, we can systematically think through where we need visibility, what parts of that visibility we should focus on, and how we should connect the dots to improve our comprehension. We can then fill in our blind spots based on the gaps in awareness we discover through the framework structure, as shown in Figure 29.

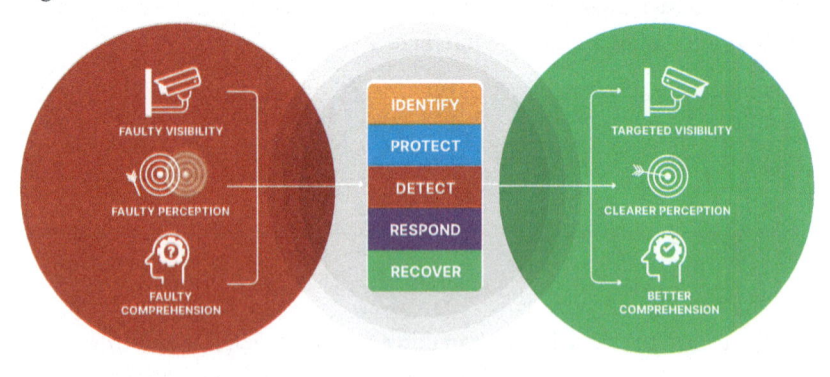

Figure 29: Leveraging Frameworks to Improve Situational Awareness

76 CYBER DEFENSE MATRIX

Structural vs. Situational Awareness

To understand how to leverage the Cyber Defense Matrix towards increasing our situational awareness, we need to first refine our terminology and improve our understanding of each of the functions of the NIST Cybersecurity Framework. The first refinement of terminology is to understand the difference between Situational Awareness and Structural Awareness. The split between these two types of awareness occurs at "boom" — the undesirable event that happens between **PROTECT** and **DETECT,** as shown in Figure 30.

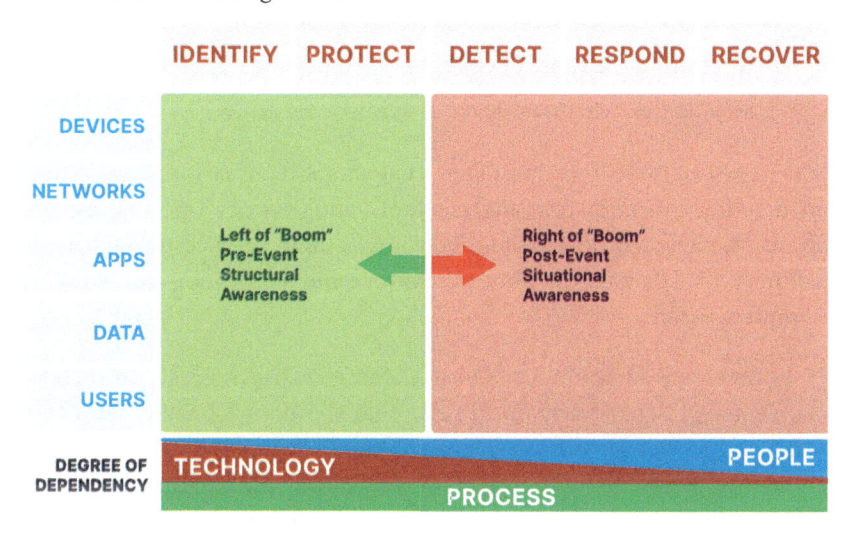

Figure 30: Left and Right of Boom – Structural vs. Situational Awareness

On the left side of boom, we need structural awareness of our environment. Most of our visibility supporting structural awareness comes from technologies that perform the functions of **IDENTIFY** and **PROTECT**, such as network firewalls, web application firewalls, and vulnerability scanners. The elements of structural awareness include the following activities:

- Understanding our valuable assets and their identity attributes
- Enumerating known structural weaknesses in those assets
- Capturing interactions with our assets
- Understanding the overall threat landscape

IMPROVING SITUATIONAL AWARENESS

We should note that activities like vulnerability scanning are on the left side of boom under the function of **IDENTIFY**. When we scan for vulnerabilities, we are looking for known structural weaknesses; thus, this activity is left of boom.

On the right side of boom, we want to establish, increase, and act on situational awareness. We want to **DETECT** if any vulnerabilities, known or unknown, have been exploited, and against which assets, by performing the following activities:

- Monitoring unexpected state or behavioral changes
- Looking for evidence of vulnerability exploitation
- Investigating the cause of changes
- Assessing the extent and severity of impacted assets

The types of **DETECT** technologies that support situational awareness include log collection and analysis tools, and Security Information and Event Management (SIEM) products. These help us slice through large volumes of data very quickly to improve our perception and support comprehension.

Note that right of boom, the Cyber Defense Matrix suggests an increasing degree of dependency on *PEOPLE* that should not be ignored. There is a limit to what *TECHNOLOGY* can do out of the box, particularly when human adversaries are deliberately trying to evade technology-centric controls. As such, regular tuning of filters and rules is an important activity that only *PEOPLE* can carry out.

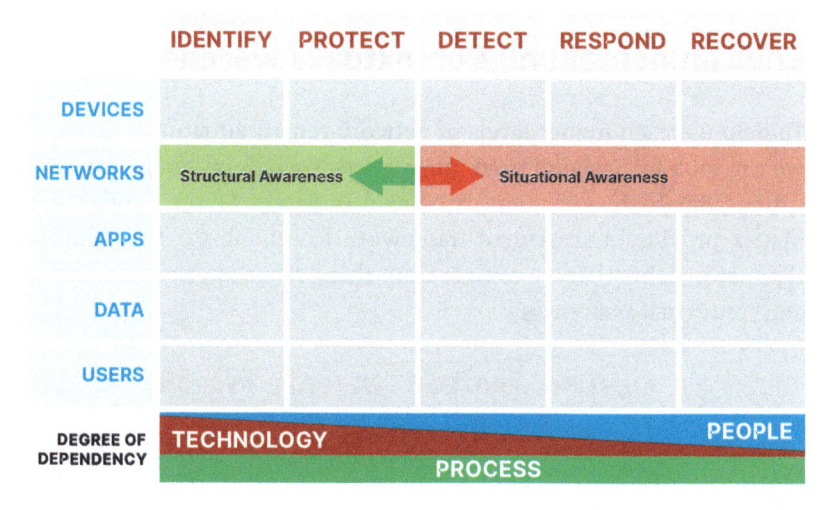

Figure 31: Network-Centric Structural and Situational Awareness

The additional dimensionality that the Cyber Defense Matrix brings to the NIST Cybersecurity Framework helps us to explicitly recognize each class of asset where we can establish structural and situational awareness. Suppose we want to focus on something happening on the network, as shown in Figure 31. On the left side of boom, we would want to establish structural awareness of our communications paths, including the following:

- Business-to-Business (B2B) links
- Virtual Private Network (VPN) connections
- Network firewall locations
- Location of possible exposures (e.g., any-any firewall rules)
- Network segments that are the most important or sensitive to the business

On the right side of boom, we want to establish network-centric situational awareness by using the visibility that we have on the left side of boom to perceive unusual changes, interactions, or communication patterns on the network. However, establishing structural and situational awareness of the network may not be enough if we are trying to find network intrusions in our environment with a high degree of precision and accuracy. We may need additional visibility to increase our level of network-centric situational awareness.

Environmental and Contextual Awareness

To help us reach higher levels of network-centric situational awareness, we can look for insights from other assets, such as our endpoints, applications, databases, and users. As shown in Figure 32, the Cyber Defense Matrix provides a structured framework to catalog the two additional types of insights that we can get from these other assets: environmental and contextual awareness.

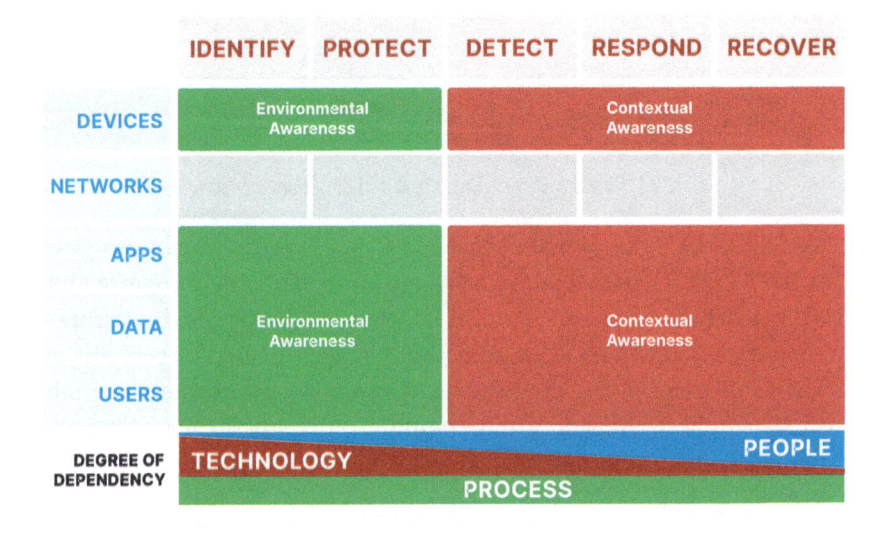

Figure 32: Network-Centric Environmental and Contextual Awareness

For **network-centric environmental awareness**, we want to know what is on the network and the state of those assets, similar to structural awareness. To that end, we want to ask the following questions:

- What devices, applications, data, and users are on the network?
- What are the upstream and downstream dependencies and interactions among those assets?
- Do those assets have weaknesses of their own which can be used to harm the network or pose danger to it?
- Are those weaknesses being monitored or addressed?

For **network-centric contextual awareness**, we want to understand what is happening around our network by observing suspicious assets that

interact with it. To that end, we want to ask the following questions:
- Has the state of devices, applications, data, or users on the network changed recently?
- What is the current behavior of those assets and how is it changing?
- What are the causes of those changes?
- Have those assets become compromised and untrustworthy?

Combining structural awareness of the network with the environmental and contextual awareness from other asset classes, this full-spectrum view provides a way to systematically and methodically elevate our situational awareness, as shown in Figure 33.

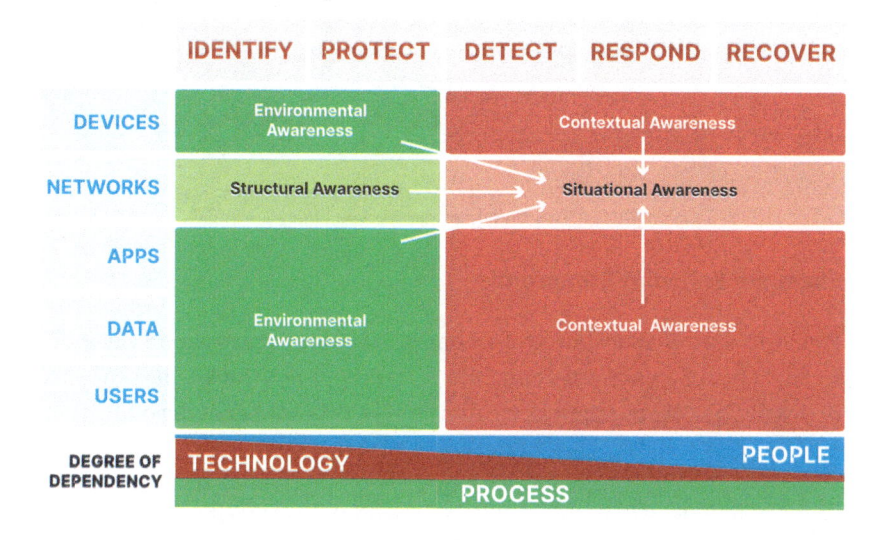

Figure 33: Full Spectrum Network-Centric Situational Awareness

These figures use the network as the center point, but we can easily shift to something else. For example, if we are looking for insider threat, we would shift the focus towards the user, as shown in Figure 34. In such a case, we would want to have structural awareness of the person, such as their position, access privileges, and vulnerabilities as discovered through background checks and phishing simulations. When we move to the right side of boom based on suspicious behaviors by the person, most insider threat programs will typically seek to achieve much higher levels of situational awareness by attaining a significant amount of environmental and contextual awareness. This will help ensure that the right decision is made about the individual before a response action is taken.

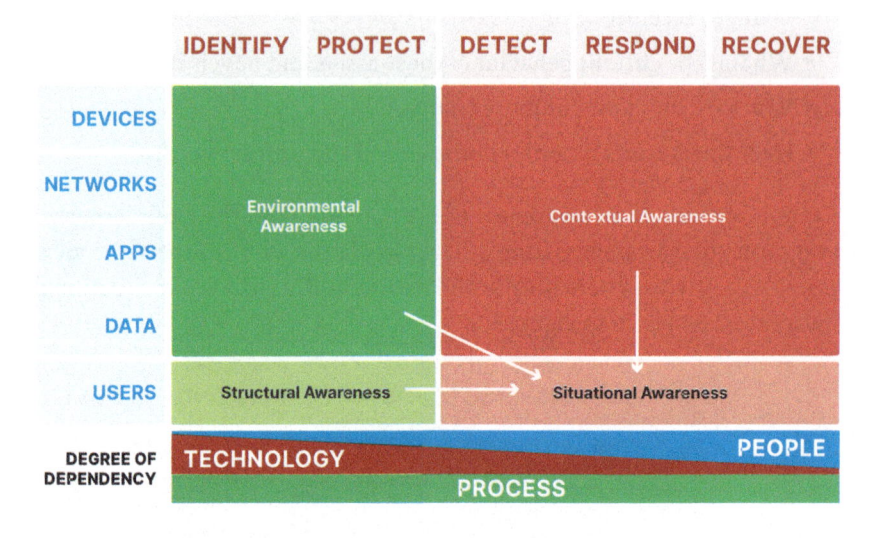

Figure 34: Full Spectrum User-Centric Situational Awareness

Example: Endpoint Compromise

Figure 35 shows an example of an endpoint compromise where these different types of awareness come into play. Stepping through the sequence of discoveries, let us suppose we find some endpoint behaving oddly (Box 1).

The first step is to gain structural awareness of that endpoint (Box 2). Let us suppose we find that the endpoint is fully patched and locked down, with 2FA enabled. There is nothing structural here to suggest why this endpoint might be acting funny.

	IDENTIFY	PROTECT	DETECT	RESPOND	RECOVER
DEVICES		❷ Fully patched, locked down endpoint, 2FA on	❶ Endpoint acting odd		
NETWORKS					
APPS					
DATA					
USERS		❸ User failed phishing simulation, no training	❹ Clicked on phish		

DEGREE OF DEPENDENCY — TECHNOLOGY / PROCESS / PEOPLE

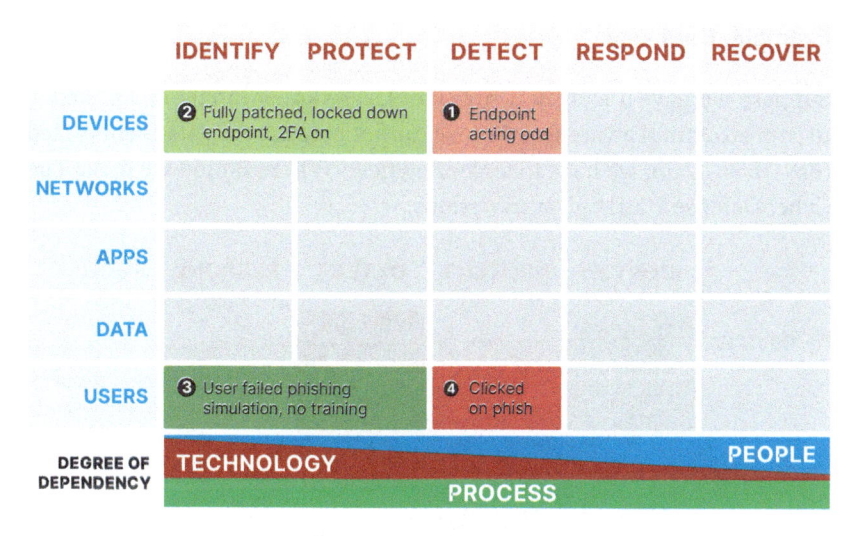

Figure 35: Endpoint Compromise Example

Our next step is to gain environmental awareness. What we may discover is that the user of that endpoint has a vulnerability (Box 3). Specifically, they failed their last phishing simulation test. Furthermore, the user has not completed their phishing training and awareness program, so they remain vulnerable.

This should prompt us to seek out contextual awareness to see if the user may have recently clicked on a real phishing email (Box 4). In the event that they did, this awareness would provide the insights needed to increase our situational awareness to understand what happened. The Cyber Defense Matrix helps us know where we need visibility, what to look for or perceive in that visibility, and how to connect the dots to comprehend what we perceive.

Let us consider another example.

Example: Data Leak

Suppose we have a DLP alert fire (Box 1), as shown in Figure 36. We try to gain structural awareness, but we cannot because the data is encrypted (Box 2). So again, we have to look elsewhere. Where should we look? The Cyber Defense Matrix gives us options.

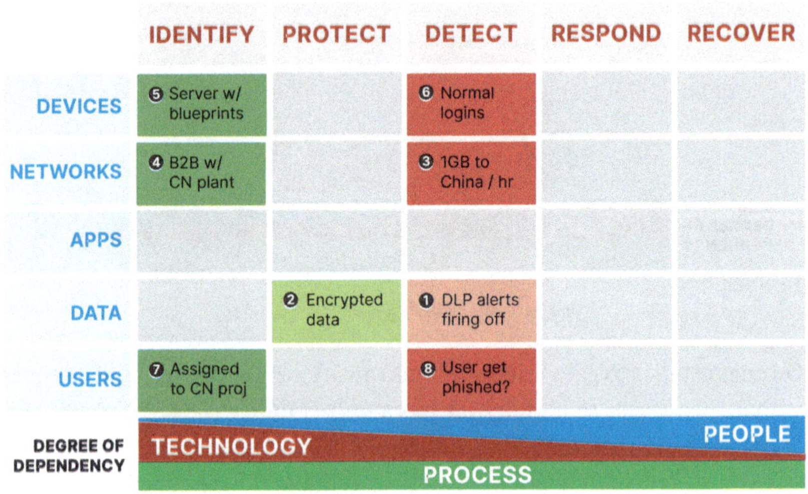

	IDENTIFY	PROTECT	DETECT	RESPOND	RECOVER
DEVICES	❺ Server w/ blueprints		❻ Normal logins		
NETWORKS	❹ B2B w/ CN plant		❸ 1GB to China / hr		
APPS					
DATA		❷ Encrypted data	❶ DLP alerts firing off		
USERS	❼ Assigned to CN proj		❽ User get phished?		

DEGREE OF DEPENDENCY: TECHNOLOGY — PROCESS — PEOPLE

Figure 36: Data Leak Example

We can get contextual awareness by looking at other events that might be happening. By looking at the network, we get contextual awareness that a machine is sending a gig of traffic to China on an hourly basis (Box 3). We can get environmental awareness by looking at this network path and seeing that a B2B connection was recently established with a Chinese manufacturing plant with whom we are doing business (Box 4).

We can seek out further environmental awareness by looking at the endpoints of that B2B connection to find a server that houses sensitive blueprints for a new product (Box 5). Getting contextual awareness for that server, we find no unusual logins or interactions (Box 6). For another confirmation check, we get more environmental awareness by seeing that the normal user of that server is an employee who is aligned to the new China project (Box 7).

With all this insight, we have higher levels of situational awareness that provide reinforcing information that this activity is probably normal business activity. However, if we were more risk averse and needed further confirmation through even higher levels of situational awareness, the framework helps us focus in on areas where we could investigate further. For example, we could get structural awareness of the user's phishing resistance levels **(USER-IDENTIFY)** and contextual awareness of the user's history to see if they have ever been successfully phished (Box 8). The Cyber Defense Matrix helps us understand what types of information are relevant in order to achieve higher levels of situational awareness.

Summary

The Cyber Defense Matrix provides a structure from which we can methodically plan our investigation as we try to comprehend what happened when an event or incident occurs. Starting with the asset class where something happened, we seek structural awareness first. How important is the asset that was affected? What are its known vulnerabilities? What is its expected behavior? What else does it normally interact with?

Then, based on what we can find out about other assets it normally interacts with, we start seeking out environmental awareness of these assets. From there, we pivot to contextual awareness of these assets. Each step of the way, we increase our situational awareness to a point where we can comfortably project what will happen next and take the appropriate courses of action. By combining these three types of awareness (structural, environmental, and contextual), we can increase our overall level of situational awareness so that we can thoroughly answer the who, what, when, where, and how questions that predictably arise when an incident occurs.

Now, during an incident, sometimes we will lack sufficient visibility in essential places. The Cyber Defense Matrix opens up options by exposing opportunities to leverage existing visibility in places that we may have not yet considered. Ultimately, if we are totally blind, then our ability to perceive, comprehend, and project will be limited. John Allspaw once observed that we need to use yesterday's incidents to inform future architectures and rules;[2] and, I will add, where we need visibility. The Cyber Defense Matrix helps us realize that we have five types of senses, and each one can provide some level of visibility or telemetry that can raise our situational awareness. Knowing is half the battle. The Cyber Defense Matrix helps us know what else may be at our disposal now, or should be at our disposal in the future.

---

2 John Allspaw, How Your Systems Keep Running Day After Day, DevOps Enterprise Summit, April 30, 2018, https://itrevolution.com/john-allspaw-how-your-systems-keep-running-day-after-day/.

CHAPTER 7

Understanding Security Handoffs

If two men on a job agree all the time, then one is useless. If they disagree all the time, then both are useless.

Assuming Responsibility

As a security practitioner, you may have wondered more than once how you came to own a particular responsibility that clearly ought to be someone else's job. One of the things that the Cyber Defense Matrix helps us understand is how the security team relates to its business partners within the organization — the partner teams that own and use the assets which the security teams are trying to secure. This chapter looks at how to parcel out the five functions of the matrix and their various subfunctions among the security team and its partner teams.

An organization where everyone expects that the security team alone will handle security, and that no one else needs to get involved, will never be very secure. To make an organization more secure, everyone must do their part. The Cyber Defense Matrix helps determine who the appropriate partner team is for any given asset class and the nature of the interaction between security and the partner team. Although different asset classes involve different partners, the handoffs mostly happen in the same place, as shown in the following figure, regardless of the asset class or its owner. As we drill down into the subfunctions, you will see that this consistency across asset classes continues to apply at more granular levels.

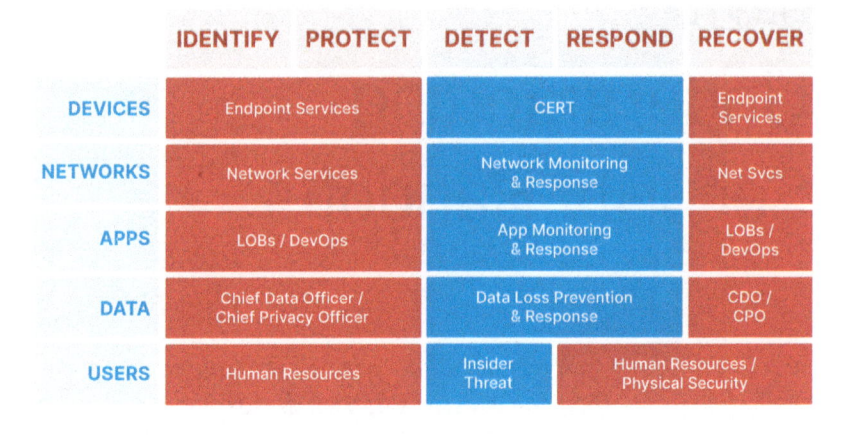

	IDENTIFY	PROTECT	DETECT	RESPOND	RECOVER
DEVICES	Endpoint Services		CERT		Endpoint Services
NETWORKS	Network Services		Network Monitoring & Response		Net Svcs
APPS	LOBs / DevOps		App Monitoring & Response		LOBs / DevOps
DATA	Chief Data Officer / Chief Privacy Officer		Data Loss Prevention & Response		CDO / CPO
USERS	Human Resources		Insider Threat	Human Resources / Physical Security	

Figure 37: Responsibility Handoffs Between Security Teams and Business Partners

At the macro level, many of the left of boom security functions and sub-functions of **IDENTIFY** and **PROTECT** are primarily the responsibility of the asset owner, who is best positioned to carry them out effectively. Only after a security incident has happened (i.e., right of boom) does the security team step in to take primary responsibility for the **DETECT** and **RESPOND** functions. Then the baton is passed back to the asset owner for **RECOVER**. Consider a physical example of commercial buildings and fires as an analogy. The primary responsibility for understanding what assets exist in the building (**IDENTIFY**) and preventing and safeguarding against fires (**PROTECT**) is the building owner's. If a fire starts, the fire department is notified (**DETECT**) and they put out the fire (**RESPOND**). The primary responsibility of returning back to normal operations (**RECOVER**) falls back to the building owner.

When this alignment is not properly executed, security practitioners feel burdened with responsibilities that they should not own. The consistent structure of the Cyber Defense Matrix helps make the case for a more logical division of responsibilities between the security team and its business partners. Aligning responsibilities this way optimizes the overall efficiency of the organization — and avoids the disconnects and the finger-pointing that misalignments cause.

One caveat: while the Cyber Defense Matrix shows what I believe may be the most efficient alignment of responsibilities, it may not be the most effective given the talent and tools that are available within an organization. There may be valid reasons to trade efficiency for effectiveness by having the security team take on the responsibilities for subfunctions that belong to the asset owner. When this happens, the security team should note this deviation with their business partner and ensure that they have the proper budget and management support to do the job well.

Handoffs in IDENTIFY

Drilling down into the **IDENTIFY** subfunctions and using a simplified RACI (Responsible, Accountable, Consulted, Informed) framework, the table below shows with greater granularity how these responsibilities should be parceled out between the security team and the asset owner.

IDENTIFY Subfunctions	Responsible / Accountable	Consulted / Informed
Inventory / Catalog / Directory / Registry	Owner	Security
Identity/Certificate/Key/Token/IP Address/DNS Mgt	Owner	Security
Prioritization / Classification	Owner	Security
Vulnerability Discovery / Attack Surface Management	Security	Owner
Threat Assessment	Security	Owner

Table 4: RACI Chart for IDENTIFY Subfunctions

Let us go into each of these subfunctions in greater detail.

Inventory

The first subfunction is that of inventory. Depending upon the asset class, we might typically call this inventory something else, such as a **DATA** catalog, **USER** directory, or **APPLICATION** registry. Regardless of asset class, it makes the most sense for the asset owner to have the primary role in inventorying their own assets. The security team is the beneficiary of that inventory. Asset owners are best positioned to perform this function, yet in many organizations, the owners often abdicate this responsibility, resulting in inaccurate and out-of-date inventories.

There are several reasons why this happens. Let us take **DEVICES**. The asset owner — the IT department — may be satisfied with a low quality inventory, one that only successfully covers 80% of their assets. They may feel that for their purposes, ignoring BYOD endpoints, orphaned devices, or decommissioned servers is fine.

For cloud assets, this is less of a problem. The users of cloud assets are constantly being billed for them, so they are incentivized to keep an accurate and complete inventory. But for on-premises assets, the owner might consider them a sunk cost and have little or no incentive to maintain a good inventory.

This problem is not unique to **DEVICES**. Consider the inventory of the **USER** asset class. Human resources (HR) is best positioned to maintain an authoritative people directory of **USERS**, but HR generally only keeps track of employees. Details about contractors, freelancers, and other users who might need access to the organization's systems are often in disparate systems, resulting in the absence of a single authoritative system of record that security can use to get details about a **USER**.

When HR, IT, or any asset owner fails to take responsibility for inventory, the security team often steps in to do so, recognizing they have much more at stake if an up-to-date, accurate inventory is not available. However, doing this work takes time and attention away from performing core security functions. In lieu of taking full responsibility for establishing an authoritative system of record for those assets, the security team should instead serve as quality control for those inventories.

To this end, the Cyber Defense Matrix provides a great framework to support quality control efforts by helping security teams discover and reconcile discrepancies in inventories across asset classes. Cyber security expert Phil Venables proposes a multi-domain approach to inventory assets, enabling reconciliation across domains.[1] He notes that most of the discrepancies we see in asset inventories are innocuous and stem from data quality issues,

---

[1] Venables, Phil, Taking Inventories to the Next Level, https://www.philvenables.com/post/taking-inventories-to-the-next-level-reconciliation-and-triangulation

race conditions among data updates, or a mismatch in reference fields. But what security really cares about are hazardous discrepancies that come from unmanaged assets. Because both types of discrepancies often look the same, it is important to diligently clean up the innocuous discrepancies to enable faster spotting of the hazardous discrepancies. As shown in the diagram below, Venables suggests connecting inventories across asset classes to find these types of reconciliation opportunities.[2]

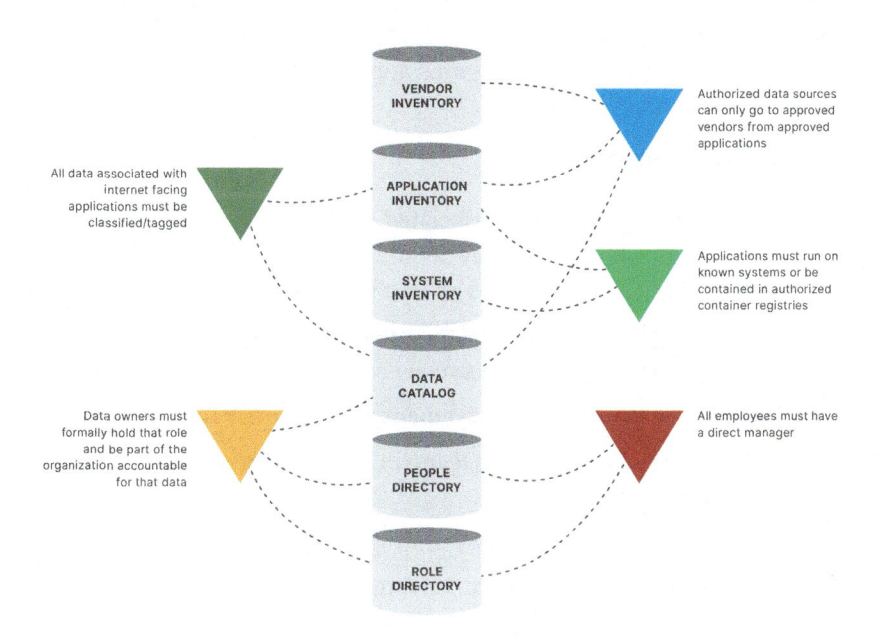

Figure 38: Incentivizing Inventory Reconciliation through Cross-Connecting Asset Classes

Ideally, security teams will want to find ways for an up-to-date inventory to provide value back to the owner. This would create a virtuous cycle that motivates asset owners to continually maintain the inventory. Venables suggests creating automated workflows that expedite approvals for certain requests (e.g., sending data to a vendor) when the associated inventories are up to date. For example, the request to send data to a vendor is automatically approved when the data is in a data catalog, the vendor is in

---

[2] I have made some slight modifications to this diagram, but you can find the original in Venable's previously cited blog post.

the approved vendor registry, and the application that is sending the data is in the application inventory.

In summary, the structure of the matrix consistently shows how the inventory function should be handled by the asset owner regardless of asset class. When this practice is not being consistently followed, the security team can use the matrix to point out these discrepancies and then devise strategies by which the owner can fulfill this responsibility. Subsequently, the security team would serve as a quality control function for reconciling inventories or creating accelerated approval paths which incentivize keeping inventories up to date.

Identity management

The notion of identity is often associated only with the **USER** asset class. However, if we think more broadly about identity, we will find that every asset has an identity of some sort and that this identity needs to be appropriately managed throughout the asset's lifecycle. For example:

- **DEVICE** ⟼ Device certificate management
- **NETWORK** ⟼ IP address management, DNS & DHCP management
- **APPLICATION** ⟼ API key management, SSL/TLS certificate management
- **DATA** ⟼ Checksums, hashes
- **USER** ⟼ Username/password, MFA tokens

The responsibility of managing these identities should consistently align with the asset owner. Security teams should not manage IP addresses; this is the job of the network team. Security teams should not manage the SSL/TLS certificates for an application; this is the job of the line of business or application development team. We will often find a discrepancy in this pattern where security teams are managing and issuing user identities (e.g., MFA tokens). If we were to follow the patterns that we see in the Cyber Defense Matrix, we could argue that it would be more efficient to have HR issue and manage those MFA tokens as they do employee badges.

Prioritization

Consider the prioritization subfunction. For the **USER** asset class, the enterprise's org chart completes this natively by providing a hierarchical structure showing who is more important. Likewise, for **APPLICATIONS**, the lines of business that use applications will identify which ones are critical since they want them protected. For the **NETWORK**, the IT department will inventory and prioritize the communication paths in and out of the environment. But for some reason, when it comes to **DATA**, security is often expected to fulfill the prioritization subfunction.

Obviously, to do this, they need guidance from the data owners about which kinds of data are most important, most sensitive, and need to be most carefully protected. Again, there is a whole class of security tools designed to do this kind of post facto labeling — for instance, by finding documents with a particular marking like "confidential" — and ensuring that they receive extra protection. But this is a hugely inefficient way of performing this function. Data should be classified by the people that create it, as part of the creation process, so that appropriate protections can be baked in from the start. Trying to prioritize and protect data later, after it has been created and stored, is inefficient and ineffective.

Again, the structure of the Cyber Defense Matrix demonstrates the logic of making the **DATA** owner — like the owners of every other asset class — responsible for its prioritization, helping us make this case.

Vulnerability discovery

The security team is responsible for finding vulnerabilities, but the asset owner needs to be consulted. That means having a two-way conversation to coordinate vulnerability scanning and the vetting of the results to score the severity properly. In many cases, the security team may discover a vulnerability but not know who the asset owner is. This increases the importance of good asset management and inventory practices so that any vulnerabilities that are discovered can be quickly communicated back to the owner for remediation.

Threat Assessment

The security team is responsible for keeping track of threat intelligence and informing the asset owners when there are elevated concerns around potential attacks. In addition, the security team is usually the driving force to conduct threat modeling exercises with the asset owner. Ideally, the asset owner should include threat modeling as a normal part of their overall design process, but in practice, this is security's responsibility to ensure that threat modeling is conducted as a part of any design effort.

Handoffs in PROTECT

Turning to **PROTECT**, there are many different subfunctions as shown in Table 5. At a high level, these activities can be grouped into patching, hardening, and setting policy.

Fixing vulnerabilities, hardening, and patching need to be done by the asset owner, who is best placed to ensure that these vital functions are performed in a way that is least disruptive to business processes.

The role of security is to set policy — to define the rules that asset owners must follow in fixing and patching, such as how quickly they have to deal with high consequence vulnerabilities. Like a regulatory agency, security will set the rules and police them, but it is up to the asset owners to actually do the work.

One common misalignment here happens with regard to the **USER** asset class. Fixing vulnerabilities in this asset class means training and security awareness. HR is responsible for all kinds of employee training, from diversity to procedures for submitting expenses to filing timesheets. They bring in subject matter experts, schedule sessions, and ensure that employees attend. Yet for some reason, in many organizations, the security team finds itself responsible for employee training in security awareness.

The correct role for security here ought to be the same as for every other asset class: setting the policy. That means defining the content and objectives of the training, and specifying how frequently it should occur. It might even mean providing SMEs to deliver parts of the training. But HR, as the asset owner, has to step up and actually be responsible for executing the security training, the same way it does for every other kind of training.

Again, the structure and consistency of the matrix help us explain why training — patching **USERS** — is the role of HR. And it helps us frame an argument that is about efficiency, the best use of resources, and the correct alignment of responsibility and accountability within our organization.

	Devices	Networks	Applications	Data	Users
Define Policy	• Device usage policy • Baselines, standards	• Network access policy	• SaaS usage policy • App Dev policy	• Classification guide	• Acceptable use policy
Repair / Correct / Patch / Remediation	• Patch	• Modify ACLs	• Fix bugs • Patch	• Encrypt • Chg permissions	• Training and awareness
Least Privileges / Control Access / Separation of Duties	• Active Directory / LDAP / RADIUS • Admin Priv Escalation	• VPN • NAC • Firewalls, VLANs	• Mutual TLS, API keys • User authentication • Dev/Test/Prod sep	• Need to know • Database activity monitoring	• IAM, Access review • SoD, Roles & Responsibilities
Credential Management	• Credential Vaulting	• Credential Vaulting • TACACS	• Credential Vaulting • API key management	• Key management system, HSMs	• Password manager • Multifactor Auth
Allow / Deny List	• Signature-based (A/V) • Behavior-based (HIPS)	• Default deny ACLs • Default allow ACLs	• Run-time application self protection	• HMACs • DLP	• Persona non grata • Full ID check
Harden	• Configuration hardening	• Close comms paths • TLS encryption	• Securing coding • Code obfuscation	• AES encryption	• Training and awareness
Least Functionality	• Unnecessary service removal	• Single packet authentication	• Microservices • Containers	• Need to know • Differential privacy	• N/A
Isolation / Containment	• Virtualization, sandboxing	• Walled garden, remediation network	• Containers, Content Disarm and Reconstr	• Rights management system	• Quarantine, jail
Audit / Log Events	• A/V, HIPS, login, and other event logs	• VPN, NAC, Firewall, DNS, DHS, DHCP logs	• Application logs • Authenication logs	• Database logs, DB activity logs	• User activity logs
Change Management	• File integrity monitoring	• Network change management tools	• Version / release control	• Integrity monitoring	• Manager 1:1s
Exploit / Attack Mitigation	• EMET, ASLR, DEP	• IPS	• Web App Firewall		

Table 5: PROTECT Subfunctions

UNDERSTANDING SECURITY HANDOFFS 95

Handoffs in DETECT and RESPOND

During the **DETECT** and **RESPOND** functions, most of the responsibilities align with the security team. With an active security incident underway, the security team will need the authority to break down doors or break glass when needed to put out the fire. Ideally, these authorities are established ahead of time so that when an incident occurs, the security team will not need to negotiate for access, but instead, will have the ability to execute its response functions without impediments.

There is one inconsistency with the special case of the **RESPOND** function for the **USER** asset class. For every other asset class, security has the **RESPOND** function. They are responsible for dealing with an intrusion. But, as I have pointed out previously, **USERS** are a unique asset class, because they are people and not machines.

With **USERS**, it is vital to distinguish between a vulnerability and a threat. A **USER** may represent a vulnerability because they demonstrate insufficient security awareness. A **USER** might even be reckless, violating security policies for innocent reasons. Again, that is a vulnerability, not a threat. Identifying a **USER** as a threat should require meeting a high standard, but once we do so, the **RESPOND** function is to remove him or her from the organization. In many cases, this will involve physically escorting someone from the premises, perhaps retrieving company devices or physical ID tokens. That often is a job for physical security, not IT security.

Handoffs in RECOVER

Lastly, when we get to the function of **RECOVER**, the security team should take a back seat and turn the reins over to the asset owner again. The security team can conduct after-action reviews and produce post-incident reports; however, the actual recovery activities are best handled by the asset owner.

Being Comfortable with Inefficiencies

Using the Cyber Defense Matrix, we can find misalignments of job functions within our organization. Furthermore, the consistent structure of the Cyber Defense Matrix provides an intuitive justification for why certain job functions might need to be realigned for efficient performance. However, not all activities that are efficiently performed are necessarily effective. In many cases, the expertise of your security team might be needed to bootstrap activities that ideally should be handled by the asset owner. Nonetheless, if you are in the situation where you are doing the job of the asset owner, knowing what is the ideal alignment should give you the insight to plan accordingly and ensure that the asset owner is eventually able to take on the responsibilities that they rightfully own.

CHAPTER 8

Investing and Rationalizing Technologies Using the Cyber Defense Matrix

*Chance favors the
connected mind.*
                          -Stephen Johnson

Exploring Investments and Startups

The Cyber Defense Matrix can be used to discover investment opportunities and new startup ideas. In 2020 and 2021, I had the privilege of working with YL Ventures, a venture capital firm focused on seed-stage Israeli cybersecurity startups, to test the matrix's value in actual investments that we made. Using the matrix, we were able to discover gaps in the marketplace and possible technical and go-to-market approaches to fill them. Whether these startups will be commercially successful is still to be seen as of this writing, but based on the feedback that I have received thus far, the methodology borne out of the Cyber Defense Matrix appears to work.

The Cyber Defense Matrix has also proven to be a useful tool for macro-level portfolio analysis. Because YL Ventures focuses exclusively on early-stage cybersecurity companies, there is a higher probability that prospective entrepreneurs will propose solutions that may seem to compete with other companies in the existing YL Ventures portfolio. While it is ultimately up to the venture capital partners and the entrepreneur to decide on whether or not a conflict of interest exists, the matrix provides a structured approach to understand where there might be potential collisions among different companies within a portfolio.

This use case also can be applied towards technology rationalization efforts. As practitioners, we often feel like we have redundant, overlapping capabilities that seem wasteful relative to our meager security budgets. The same use case which helps discover marketplace gaps can also be used to understand where we have opportunities to reduce our technology footprint and inform our technology refresh strategies.

Finding Gaps

Suppose we could characterize a universal set of security capabilities using the letters A through Z, as shown in Figure 39. For any given asset class, it may be difficult to find a set of vendors that can collectively offer a complete set from A–Z. This is true for at least three reasons.

	IDENTIFY	PROTECT	DETECT	RESPOND	RECOVER
DEVICES	A- -C-D-	F-G-H-I-J	K-L-M- -	P- -R-S-T	-V- -X- -
NETWORKS	A-B-C- -E	F- -H-I-	- -M-N-	- - -S-T	- -W- - -Z
APPS	A- -C-D-E	-G- -I-J	-L-M- -	P- -R- -T	U- - - -Y-
DATA	A-B-C-D-	F- -H-I-	- -M- -O	P- - -S-	-V-W-X- -Z
USERS	- -C- -E	-G-H- -J	- -M-N-O	- - - -	U-V- -X-Y-

DEGREE OF DEPENDENCY: TECHNOLOGY — PROCESS — PEOPLE

Figure 39: The Universal Set of Security Capabilities (and Gaps)

- Reason #1: There is a tendency for vendors to offer point solutions that address only a specific problem (e.g., Vendor 1 provides capability D for **DEVICES** but not capability E).
- Reason #2: We may recognize the need for a capability (e.g., capability

D for **APPLICATIONS**), and there may even be some localized / in-house tooling to address it, but perhaps due to lack of commercial viability (e.g., inability to scale, niche buyer segment, etc.), we have not seen it become commercially available.

- Reason #3: We may simply not be aware that a capability gap exists (e.g., capability Q), and it is waiting to be discovered or invented, usually because a new class of technology has emerged (e.g., quantum computing).

Gaps from Reason #3 cannot be discovered through the Cyber Defense Matrix, but the matrix does help to uncover gaps emerging from Reasons #1 and #2.

Rationalizing Technologies

Gaps discovered through Reason #1 point to possible merger or acquisition (M&A) opportunities for larger companies that need to fill holes in their portfolio with point products to be able to offer a complete set of capabilities from A–Z. For example, if you are a defender or a large vendor that currently has F, G, I, and J within your portfolio of **DEVICE-PROTECT** capabilities, it stands to reason that you should be on the lookout for a vendor (most likely a startup) that offers capability H.

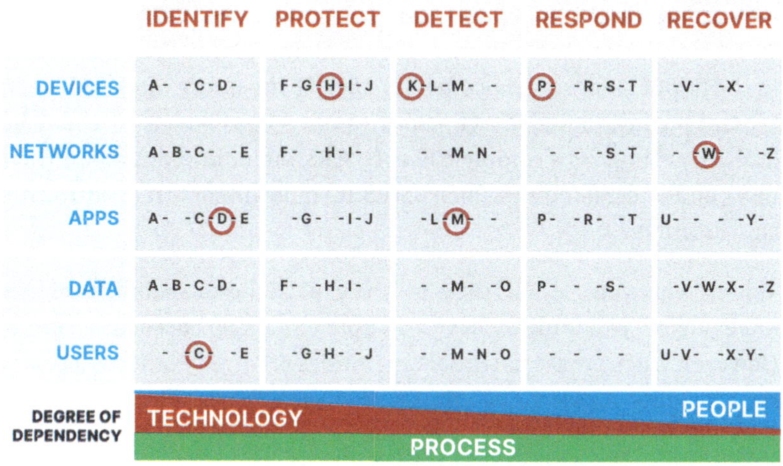

	IDENTIFY	PROTECT	DETECT	RESPOND	RECOVER
DEVICES	A- -C-D-	F-G-(H)-I-J	(K)-L-M- -	(P)- -R-S-T	-V- -X- -
NETWORKS	A-B-C- -E	F- -H-I-	- -M-N-	- - -S-T	- (W)- - -Z
APPS	A- -C-(D)-E	-G- -I-J	-L-(M)- -	P- -R- -T	U- - - -Y-
DATA	A-B-C-D-	F- -H-I-	- -M- -O	P- - -S-	-V-W-X- -Z
USERS	- (C)- -E	-G-H- -J	- -M-N-O	- - - -	U-V- -X-Y-

DEGREE OF DEPENDENCY: TECHNOLOGY ← → PEOPLE / PROCESS

Figure 40: Identifying Point Products and Potential Overlaps

From a buyer standpoint, noting where we have point products, as shown in the circled letters in Figure 40, also helps us understand where there is a strong potential for overlapping capabilities over time. As larger vendors build or acquire the capabilities found in point products, we can use the Cyber Defense Matrix to see and anticipate where overlaps may occur to help with our technology rationalization efforts. For example, if we have a point product vendor providing capability D for **APPLICATION-IDENTIFY**, it is highly likely that one of our larger vendors for capabilities A, C, and/or E will build or acquire a similar capability D, thus creating an overlap.

If limited budgets force us to let go of a vendor, having this macro view can help with the technology rationalization exercise to determine which exact capabilities will be lost. We can then work with our remaining vendors to include those missing capabilities in their future roadmap.

Finding Investment Opportunities

Gaps discovered through Reason #2, where we see parallel capabilities across other asset classes, point to possible startup investment opportunities. Let us suppose that capability A is "inventory or visibility," and capability B is "prioritization or classification." We would then expect to see a capability to do A and B across all five asset classes. However, we often do not find vendors that offer these capabilities across all five asset classes. For example, in the **DATA** asset class, we can easily find **DATA** Inventory and **DATA** Classification tools. But are there similar tools for other asset classes? In the **DEVICE** asset class, we often find **DEVICE** Inventory tools, but we may be hard pressed to find **DEVICE** Prioritization tools. Likewise, we may have difficulty capturing a complete list of **USERS**, including contractors (inventory), and understanding which **USERS** are most important (prioritization/classification).

When "what you see is all there is," the gaps in these capabilities are often not evident.[1] We think we have a complete set of our needed capabilities. However, when we see potential capabilities relative to other asset classes, the capability gaps become immediately apparent, as shown in Figure 39.

---

1 This phrase is borrowed from Nobel Laureate Daniel Kahneman's book, Thinking, Fast and Slow. It is a must read for everyone who practices risk management.: https://us.macmillan.com/books/9780374533557/thinkingfastandslow.

Some of these gaps become particularly obvious when a new subclass of asset (e.g., IoT, APIs, 5G, etc.) emerges in our digital environment. With each new subclass of asset, we often follow the same pattern of capability needs and development. For example, when IoT or SCADA security became a major concern, the first capabilities were focused on inventorying and visibility. Over time, other capabilities become available in the marketplace.

The pattern of capabilities evident in the Cyber Defense Matrix makes it easy for startup founders and investors to anticipate the need for a complete set of these capabilities. This does not necessarily mean that building capabilities will yield a viable company. The buyer market might not be ready. The adjacent technologies might not be available. There could be a wide range of reasons why the timing is not right. However, one can look for similar patterns in how new capabilities were brought to market by successful entrants. For example, what was the go-to-market strategy for companies whose initial focus was on visibility? How did successful companies expand to cover other needed capabilities? The parallels that the Cyber Defense Matrix reveals can help founders understand a path towards addressing these gaps in a way that increases the chances of commercially successful outcomes.

As for actually solving these gaps, cybersecurity is a wickedly hard problem, and solutions often elude us. However, David Epstein, in his book *Range*, observes that wickedly hard problems in economics, science, and society are often solved by outsiders, a phenomenon he attributes to the outsider's ability to make divergent analogies. When considering possible solutions to a hard problem, these analogies help in surfacing non-obvious solutions. Often, the main challenge for this approach is coming up with the most relevant analogy.

Fortunately, the Cyber Defense Matrix provides a ready-made structure to tie in relevant analogies. Solutions for one asset class provides a pointer to possible solutions in other asset classes, or even within the same asset class. For example, we saw an evolution from firewalls attempting to control inbound network traffic to "next generation" firewalls with controls for outbound network traffic. This pattern is currently repeating itself in email security, where secure email gateways initially addressed inbound email traffic and "next generation" secure email gateways are now providing controls for outbound email traffic. What would be the equivalent analogy for endpoint devices, application APIs, or datastores? Discovering these parallels in other asset classes offers ample opportunities to build new categories of capabilities that could be brought to market. Although this discovery process might not yield ideas that are commercially viable at the present time, the Cyber Defense Matrix provides a foundation for understanding how to think through a wide range of potentially needed capabilities using a systematic process.

Characterizing Vendor Movements

The cybersecurity landscape is not static, and we should not expect vendors operating within it to remain stationary. New attack surfaces, novel attacker techniques, and constant integration challenges conspire to keep vendors and practitioners on our toes to keep up with the constantly evolving landscape. With this in mind, the Cyber Defense Matrix can be adapted to reflect the typical motions that we see in the marketplace as shown in Figure 41.

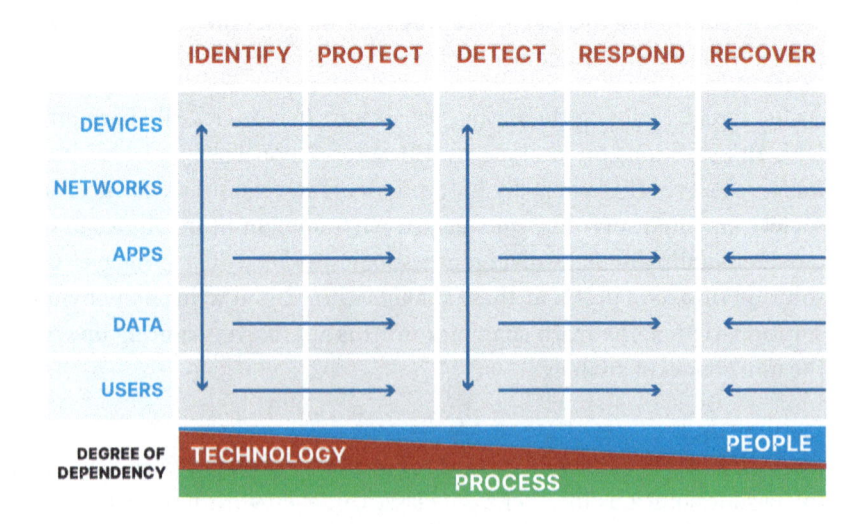

Figure 41: Vendor Movements across the Cybersecurity Landscape

Most movements are horizontal across a given asset class. On the left side of boom (**IDENTIFY** and **PROTECT**), a vendor usually starts with an **IDENTIFY** capability to provide visibility, or structural awareness, into a given asset environment. This structural awareness focuses on the state of the asset, often including some combination of inventorying assets, understanding the relationships among them, and enumerating their vulnerabilities, configuration flaws, and attack surfaces. However, getting structural awareness is usually not sufficient. The next demand from the customer is "what can I do about what I am seeing?" This often prompts vendors to develop capabilities to directly address deficiencies, which enables that vendor to claim that they cover the function of **PROTECT**. The tooling to support **PROTECT** is often distinct from the technologies that provide structural awareness, making vendors reluctant to commit resources to building such capabilities. Instead, they often choose to enable integrations with other vendors (which means that the original vendor is technically not doing **PROTECT**) or perform an acquisition that brings the two adjacent capabilities together.

On the left side of boom, vendors are able to extend their capabilities vertically when the underlying processes are similar. For example, the activity of vulnerability management is often similar across all asset classes. The process involves collecting the results of vulnerability scans, prioritizing the findings, assigning owners to the findings, and keeping on top of the owners to ensure that the vulnerability is addressed. Most vendors operating in this space often focus on one asset class. Some extend vertically and address two or three asset classes (most often device, network, and application-centric vulnerabilities). I have not seen a vendor address vulnerability management across all five types of assets. This absence of capability means that we must try to link these assets ourselves; until we do, we will have gaps in our holistic understanding of vulnerabilities in our environment. As mentioned in Chapter 1, a bug-free application running on a secure device within a highly segmented network with fully encrypted data might still get compromised if these assets are administered by someone who clicks on every phishing email they get. The vertical extension of capabilities does not typically happen in **PROTECT** primarily because the defensive mechanisms are usually specific and tailored to each asset class.[2]

---

2 Even when the defensive mechanism is the same, e.g., a host-based firewall and a network firewall, the technology for the management plane is different.

On the right side of boom, a vendor often starts with telemetry for an asset environment and is able to make sense of that telemetry to perform the function of **DETECT**.[3] After an event passes through various filters, thresholds, and rules, an alert might be generated that points to an event that demands further investigation to determine if a **RESPOND** action is warranted. For many situations right of boom, we usually see too many false positives to feel comfortable in automating the immediate execution of a **RESPOND** action when an alert fires.[4] However, it is natural for vendors to offer (and buyers to ask for) a **RESPOND** capability that is tightly coupled with their **DETECT** capability.

As with the interplay between capabilities that support **IDENTIFY** and **PROTECT**, capabilities supporting **RESPOND** often couple tightly with **DETECT**-oriented capabilities to support incident response. This natural adjacency between analytic tools **(DETECT)** and security orchestration, automation, and response (SOAR) tools **(RESPOND)** tend to drive integrations and acquisitions among vendors servicing the **DETECT** and **RESPOND** functions.

The interpretation and analysis of telemetry uses techniques and analyst know-how that is specific to the asset class. Understanding how a device behaves when it is compromised is very different from understanding how a person behaves when they become compromised and turn into an insider threat. As such, analyst-centric capabilities do not easily pattern match vertically across asset classes. However, on the right side of boom, vendors can extend their capabilities vertically when the underlying technologies are similar.

For example, the function of **DETECT** often requires the use of a logging and analysis platform. The underlying technology that supports logging and analysis can be relatively generic and applicable across all asset classes. What differentiates **DETECT** capabilities is the know-how that is

---

3 This telemetry is sometimes also characterized as "visibility," but in this case, the visibility relates to event information (what is happening) whereas the visibility left of boom is about state information (what it is). I try to avoid using "visibility" for this reason. It is often used too loosely without further clarification and creates confusion around whether the visibility a vendor is claiming supports structural awareness or situational awareness.

4 When an alert's false positive rate becomes negligible and the RESPOND action produces a tightly scoped, deterministic outcome, this usually transitions from being reactive (right of boom) to proactive (left of boom).

applied on the telemetry. This differentiation is exemplified by technologies like endpoint detection and response (EDR), network detection and response (NDR), user behavior analytics (UBA), and user/entity behavior analytics (UEBA). It also explains the emergence of a converged capability — extended detection and response (XDR) — that leverages a common underlying technology to satisfy detection use cases across multiple asset classes.[5] We should note that the current generation of XDR products does not cover application detection and response (ADR) and data detection and response (DDR). However, startups have recently emerged that tackle these particular use cases. It should not be long before these two remaining asset classes are incorporated into the XDR paradigm.

Vendors that focus on the **RECOVER** function tend to remain focused on that function and remain within their respective asset classes. If they do expand, it is primarily horizontally within the asset class that they start in, and usually to support **IDENTIFY** and **PROTECT** use cases. For example, in the **DATA** asset class, it is not uncommon to find data backup vendors that also leverage those backups to conduct data inventory and classification, which are subfunctions under **IDENTIFY**.

Lastly, I should note that smaller startup vendors do not typically expand both horizontally and vertically at the same time, as that would significantly dilute the attention and focus of the startup. On the other hand, larger vendors do eventually end up expanding both horizontally and vertically, sometimes through organic capability development but most often through acquisitions. If they choose to expand through acquisitions, the pattern of acquisitions should follow a methodical path that follows the natural adjacencies of capabilities in their existing portfolio.

Portfolio Analysis for Investors and Practitioners

With an understanding of how vendors might expand their capabilities over time, we can analyze investment opportunities to determine if potential conflicts of interest might arise. From a practitioner standpoint,

---

5 The term "extended" poses the same problem as "cyber" in that the term is not specific to any asset class. This causes confusion around what exactly is meant by "extended." Using the Cyber Defense Matrix, when we define "extended" to mean all the asset classes, we can quickly spot the gaps in how "extended" is defined by vendors today.

this analysis also supports technology rationalization efforts to pinpoint potential capability overlaps. Removing technologies from our portfolio is often harder than adding new ones. This analysis may help in making a case for retiring certain technologies over a longer period of time as we anticipate these overlaps.

Using the diagram in Figure 42, suppose our portfolio currently consists of products A, C, F, and H. If we were considering investing in or buying products B, D, E, and G, which ones might eventually result in a conflict of interest or overlapping capabilities? Anticipating where a vendor might likely expand their capabilities helps us plan for these overlaps.

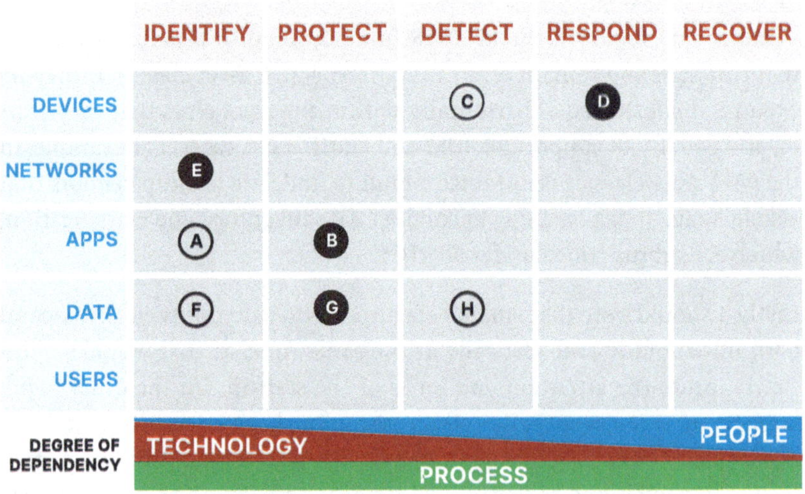

Figure 42: Mapping Current and Potential Portfolio Capabilities

When we can see these companies in relation to their expected expansion over time, we can see where there may be collisions that could create overlaps or conflicts of interest. Suppose vendors A, C, and G plan to expand horizontally, while vendors E, F, and H plan to expand vertically, as shown in Figure 43.

With these expected expansion of capabilities, we can then understand how to interpret the following three potential overlap scenarios:

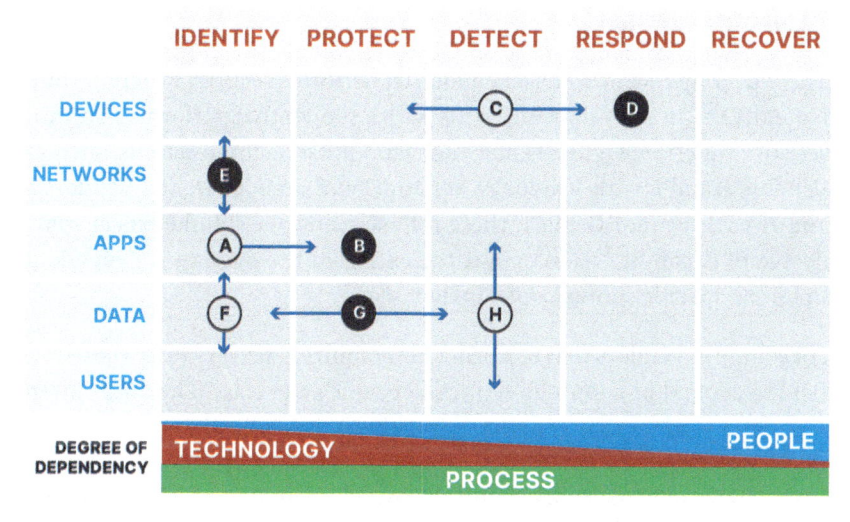

Figure 43: Anticipating Future Portfolio Conflicts

No overlaps

This should be fairly self evident, but there are no overlaps among different vendors that operate horizontally on different asset classes (e.g., Vendors A, C, and G) or vertically on different functions (e.g., Vendors E and H). An example of non-overlapping vendors on the horizontal plane would be one that does application vulnerability discovery (i.e., **APPLICATION-IDENTIFY**; e.g., static application security testing) and another that does data vulnerability discovery (i.e., **DATA-IDENTIFY**; e.g., open file share discovery).

Vendors that do not overlap can be highly complementary, but they often engage different buyer segments in larger organizations. For example, capabilities that focus on **IDENTIFY** and **PROTECT** are generally bought by those in security engineering, whereas capabilities that focus on **DETECT**, **RESPOND**, and **RECOVER** are generally bought by those in security operations. Similarly, products that address endpoint security issues are often purchased by different buyers than those that handle application security issues.[6]

---

6 These differences in buyers means a divided focus for sales and engineering, which is why it would be difficult for a smaller startup to expand in two directions at once.

INVESTING AND RATIONALIZING TECHNOLOGIES 109

Tangential overlaps

These overlaps occur when a vendor that performs security functions horizontally for an asset class intersects with a vendor that performs a vertical security function. Figure 43 shows several possible intersections, such as Vendors E and F with Vendor A; Vendor C with Vendor H; and Vendors F and H with Vendor G. With these intersections, we should expect some degree of capability overlap, but the extent of the overlap is dependent upon the specific subfunctions that overlap.

For example, vendors that perform vulnerability discovery (e.g., Vendor A) tend to expand horizontally within a specific asset class. They may intersect with vendors that support vulnerability prioritization and management (e.g., Vendor F), which tend to expand vertically. Vendors A and F may overlap in conducting vulnerability prioritization, but Vendor A should provide localized insights on prioritization (e.g., which application-oriented vulnerabilities are most severe), whereas Vendor F should provide global insights on prioritization (e.g., among all vulnerabilities across all asset classes, which vulnerabilities are most severe). Although these could be considered overlapping and redundant, we will often find that those vendors that specialize in a given asset class (i.e., operating horizontally) provide a greater depth of analysis compared to those vendors that provide breadth by transcending any given asset class (i.e., operating vertically). However, vendors that operate vertically tend to provide a more holistic view and integrate across multiple technology stacks.

An example of this playing out might look like the intersection between Vendors C and H. Let us suppose Vendor C sells endpoint detection and response (EDR) capabilities, whereas Vendor H sells extended detection and response (XDR) capabilities. We should expect depth from Vendor C, providing a deeper level of analysis to discover endpoint intrusions. Vendor C is also likely to deploy an endpoint agent. On the other hand, we should expect breadth from Vendor H. Vendor H will likely not require a dedicated endpoint agent, but could leverage Vendor C's agent (or one of Vendor C's direct competitors) to collect the necessary data on the endpoint. Vendor H should pull data from other asset classes (e.g., databases, networks, applications, etc.) to provide a more comprehensive view of an intrusion to track an attacker as it moves across the various assets of an organization. Although it may seem Vendor C and H could result in an overlap, they should treat each other, at worst, as frenemies, and at best, complementary partners that provide an equal balance of breadth and depth.

Direct overlaps

Overlapping capabilities may occur when vendors are operating horizontally in the same asset class (e.g., Vendors A and B or Vendors C and D), or vertically in the same function (e.g., Vendors E and F). In these situations, collisions are more likely, resulting in a similar set of features available from both vendors. These vendors will generally see themselves as directly competitive.

However, an exception to this arises in situations where the solutions are geared towards different types of assets within an asset class, or towards different subfunctions within a function. For example, in the **DEVICE** asset class, security products for workstations will be very different from security products for Internet of things (IoT) or industrial control systems (ICS). Similarly, in the **APPLICATION** asset class, we may see a variety of security products supporting different application languages (e.g., C++, .NET, Java) or different application types (e.g., mobile, desktop, web). In these circumstances, although two vendors may appear to overlap within a given asset class, they may not actually be overlapping in the short term or even in the medium term. They may find themselves eventually competing in the long term, but if these distinct capabilities are in currently demand, practitioners will not wait until the vendors converge over the long term to acquire what may seem like overlapping capabilities.

Choosing the Right Kinds of Overlaps

There are other ways to leverage the Cyber Defense Matrix to understand the movements of vendors to anticipate and avoid unnecessary collisions. Competition is great for fueling innovation and lowering prices in the macro economy, but we want to limit, or at least be deliberate, when we introduce overlapping capabilities within our own portfolios. In larger organizations, there are many practitioners that joke about having "bought one of everything" or practicing "expense in depth." While many of these tools may indeed be unnecessary, some overlap may be desired. This overlap can serve as a quality control function, or as a part of having additional redundancies in security controls for critical systems.

Overlaps are not all bad, but we should seek to know when and where they occur so that we plan accordingly.

CHAPTER 9

Dealing with the Latest Security Buzzwords

*Jargon live in the swamps. They feed on attention.
If they cannot get that,
they'll settle for fear and confusion.*
— Carlos Bueno

Staying Relevant Amid Change

In some ways, a cybersecurity model like the Cyber Defense Matrix resembles a scientific theory or paradigm. It should be able to explain the world as we know it right now while also accommodating and explaining new developments or discoveries. When a new discovery occurs or a new technology emerges, the scientific theory should be able to show how these novel developments are explained by the theory without contradictions. Furthermore, scientific theories should be able to predict new discoveries.

In the same way, the Cyber Defense Matrix needs to demonstrate its continuing relevance as new security technologies, architectures, and approaches arrive. It needs to show how those new products, capabilities, and designs still fit into the model the matrix provides. The matrix needs to prove its continuing utility by showing how it adds explanatory value to new things that appear in security without changing the core structure or foundations of the matrix. I have already covered the predictive power of the Cyber Defense Matrix in Chapter 8. In this chapter, I will cover some of the buzzwords that we often encounter in the marketplace and show how the Cyber Defense Matrix can explain these terms and shed new insights on how to see these buzzwords in a broader context.

Zero Trust and Secure Access Service Edge

To understand how Zero Trust and Secure Access Service Edge (SASE) map to the Cyber Defense Matrix, it helps to understand how to depict the old model of access before we started embracing zero trust principles.

In the past, we secured our enterprise at the network perimeter. Like a castle, our assets were surrounded by a moat that created a trust boundary. As shown in Figure 44, access inside to assets within this trust boundary (the TO column) was controlled through a network-centric gateway, such as a firewall or virtual private network connection. Once inside this trust boundary, transitive trust is implicitly granted, giving those inside the authorization (AuthZ) to reach all other assets within the trust boundary.

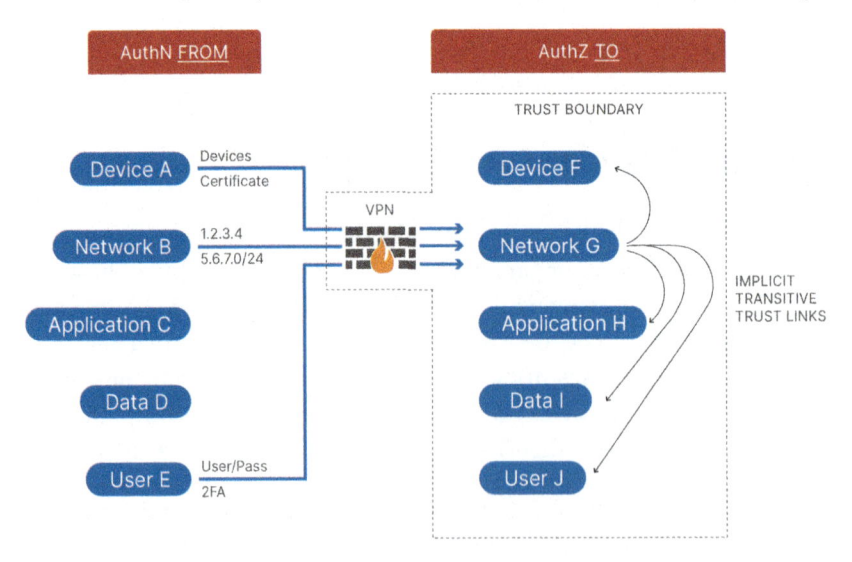

Figure 44: Network Perimeter-Based Access Model

To gain initial access, we had to present credentials for authentication (AuthN) using identity attributes associated with the requesting entity (the FROM column). This included robust credentials such as device certificates and two factor authentication, but also weak identity credentials such as username and password or an IP address within specific ranges.

To map this architecture to the Cyber Security Matrix, we can look at each

114 CYBER DEFENSE MATRIX

of the five asset classes as both a requester and a resource. For each asset class, there is a "FROM" and a "TO." The requester represents the FROM part of the equation, and the resource that the requester is trying to get represents the "TO" portion.

We can also look at this from the standpoint of the difference between authentication and authorization. For authentication, or AuthN, we want to get FROM the entity some kind of proof of identity to ensure that it is actually who or what it says it is. For authorization, or AuthZ, we want to control what that entity can access, ensuring it can only get TO the resources it is entitled to use.

Identity attributes from **DEVICE** A, **NETWORK** B, or **USER** E are presented to **NETWORK** G. If the identity attributes used for authentication are sufficiently trustworthy, then these assets are granted access to **NETWORK** G, from which those assets are implicitly given authorization to reach other assets within the trust boundary (**DEVICE** F, **APPLICATION** H, **DATA** I, and **USER** J).

As we have discovered through lessons learned from past security breaches, the assumptions underlying this security model are flawed. The network perimeter has certainly never been impenetrable. Attackers have repeatedly shown their cleverness in being able to get through. The porous nature of the perimeter also undermines the assumption that anyone inside it can be considered trustworthy. Once inside the network perimeter, attackers face an unsegmented environment in which they can move laterally with relative ease. Worse, the growth of web-based and cloud services, mobile devices, and remote working has also undermined the central assumption that the organization's key assets are behind the perimeter in the first place. Unfortunately, much of contemporary security architecture is designed based on these assumptions. If they are faulty, we need to redesign accordingly with a new set of assumptions — and a new set of foundations for what constitutes trustworthiness.

Zero Trust design principles advocate that trustworthiness should not be automatically assumed when it comes to granting access to any resource. Instead, each resource has its own trust boundary, as shown in Figure 45. Network access is not dependent on network locality but rather upon the presentation of strongly bound identity assertions, usually in the form of device certificates, username and password, and a 2FA token. The identity assertions from both **DEVICE** A and **USER** E are processed by an access proxy before being granted access to **NETWORK** G.

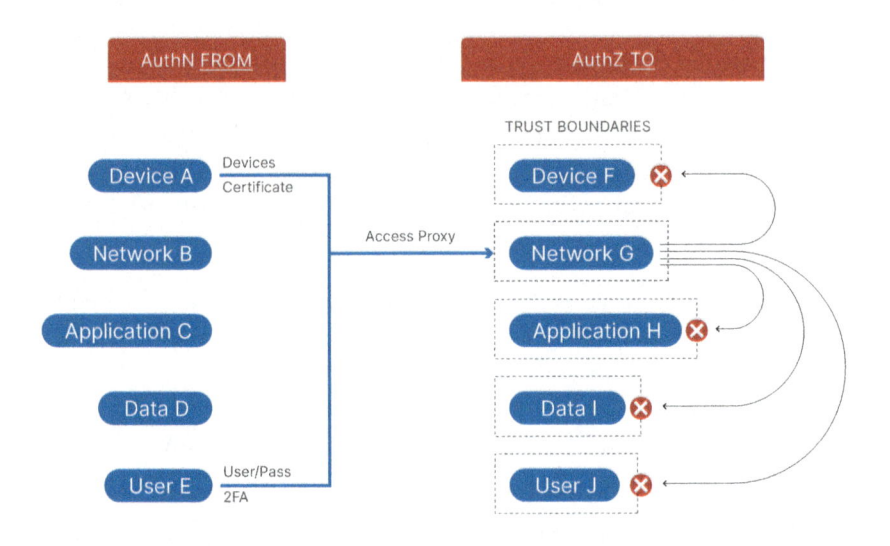

Figure 45: Zero Trust Network Access

In a Zero Trust design, we never implicitly trust the entity requesting access. For **USERS**, we want to verify the trustworthiness of the **USER** by examining three types of attributes about the **USER**:

- **Structural attributes**: E.g., username/password, fingerprint, date of birth, assigned multifactor tokens.
- **Environmental attributes**: E.g., when and from where they are logging in, what their job is, what projects they are working on, etc.
- **Behavioral attributes**: E.g., data about where they go in the **NETWORK** and what **APPLICATIONS** they try to access.

Using these attributes, we can verify a **USER's** identity and assign it a trustworthiness score. The structural attributes about the **USER** map to **USER-IDENTIFY**. Behavioral and environmental attributes are powerful additions to the verification armory. Structural characteristics can often be falsified (via account takeover) or subverted (in the case with insider threat). In the event that this happens, behavioral or environmental attributes can still give defenders important clues about a given identity.

An identity may appear trustworthy based on its structural attributes but its environmental and behavioral attributes may suggest that it cannot be trusted at all. Is the employee logging in during work hours or in the middle of the night? Is their IP address in Northern Virginia or Shanghai or is it a Tor exit node? More importantly, these attributes can be monitored continuously to ensure that context or behaviors do not change in a way that decreases trustworthiness. Is a **USER** identity suddenly downloading much larger numbers of files than usual? Are they accessing **DATA** they have never used before?

In some situations, we may want to go further and not just verify the **USER**, but also the **DEVICE** that they are using. We can similarly gather attributes about the **DEVICE** to determine its trustworthiness. Here are examples of each.

- **Structural attributes**: a unique identifier like a PKI certificate or IMEI number, the patch level for the operating system, its configuration and security posture. These all map to **DEVICE-IDENTIFY**.

- **Environmental attributes**: information about what other networks the device is connected with, what software it is running, etc.

- **Behavioral attributes**: who it is communicating with and what data it is sending.

To verify the identity of an entity and its associated trustworthiness, we can use an access proxy that can be dynamically configured to accept various identity attributes FROM the requesting entity and use those attributes to make real-time decisions about the resources it should have access TO. Any subsequent access to other resources should require its own explicit trust attestation. When we do this for access to **NETWORK**-centric resources, the current industry solution category is called Zero Trust Network Access (ZTNA) or Software Defined Perimeter (SDP). The access proxy that grants zero trust access to the **NETWORK** maps to **NETWORK-PROTECT** on the Cyber Defense Matrix.[1]

---

1 There is a danger in suggesting that there is such a thing as a "Zero Trust" product. Zero trust is an architectural design principle and not a product. Components of that architecture may be a product but they are not exclusively associated with zero trust designs. One can operate an access proxy without adhering to zero trust design principles.

Access to other types of resources represent other forms of Zero Trust access. For example, Zero Trust Application Access (ZTAA) leverages the same set of identity assertions but for access to **APPLICATIONS** as shown in Figure 46. These applications can be either custom in-house built applications or SaaS applications with the ability to support restricted access to enterprise accounts from only the zero trust access proxy. Since this type of access proxy is controlling access to **APPLICATIONS**, it would map to **APPLICATION-PROTECT**.

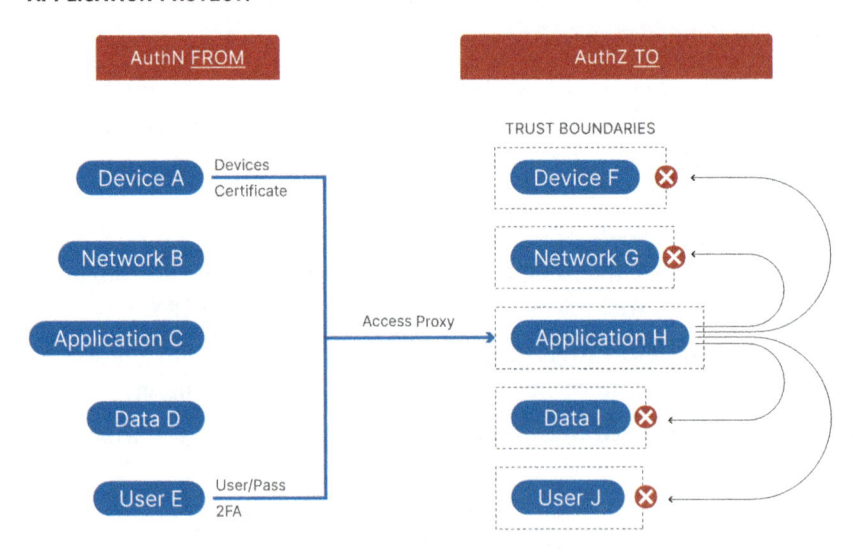

Figure 46: Zero Trust Application Access

This model can be easily extended to represent other forms of Zero Trust access to other asset classes. Figure 47 shows how it might manifest for **DEVICE**-centric resources, usually through protocols such as secure shell (SSH) or remote desktop protocol (RDP) or through a remote browser isolation (RBI) solution. Because these **DEVICE** assets often serve as a gateway to other resources, they may be allowed to reach into other trust boundaries, but to adhere to the zero trust mindset, these trusted links should be explicitly defined. The zero trust access proxy for **DEVICE**-centric assets would map to **DEVICE-PROTECT**.

You may have noticed that depending upon the use case, I have mapped the zero trust access proxy to three boxes on the Cyber Defense Matrix (**DEVICE-PROTECT**, **NETWORK-PROTECT**, and **APPLICATION-PROTECT**). We can naturally expect a **DATA**-centric access proxy to emerge in the market and that is what we now see through emerging market categories such as Data Access Security Brokers or Secure Data Access Platforms.[2]

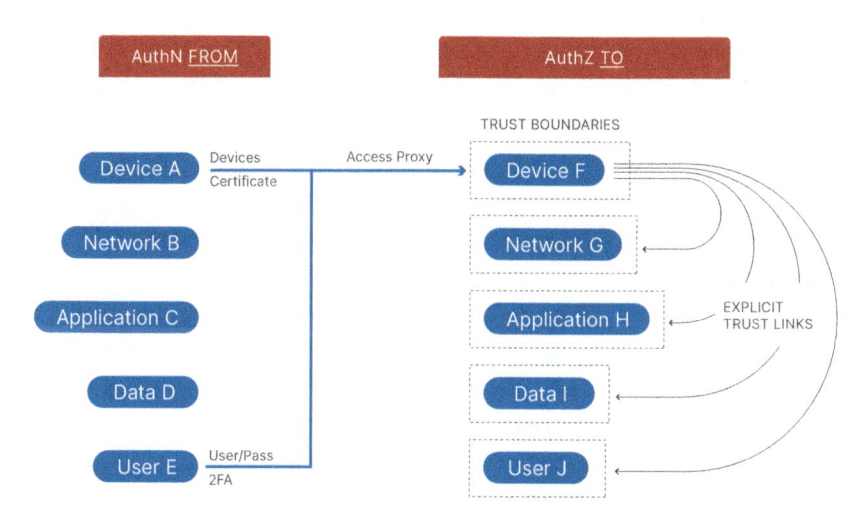

Figure 47: Zero Trust Device Access

I believe this mapping to multiple boxes defining how access to resources is protected reflects what is also happening with the evolution of Secure Access Service Edge (SASE), and specifically Security Service Edge (SSE), which is a framework that describes a way to securely connect one set of assets to another. SASE and SSE describe the convergence of multiple access mechanisms to unify visibility, improve the user experience, and simplify policy and management.

However, I believe that the Cyber Defense Matrix offers an even broader way of thinking about how access can be managed and how trustworthiness can be established. Most vendors in the market that offer converged solutions only manage access to a subset of asset classes defined by the Cyber Defense Matrix (usually just **NETWORK** and **APPLICATIONS**). We will likely see future offerings that also encompass converged access for the **DEVICE** and **DATA** asset classes.

---

2 An access proxy is not needed for **USER-PROTECT**, though if one is needed, a good executive assistant does the job well.

In addition, there are other ways to establish higher levels of trustworthiness on the identity side of the zero trust equation. When we hear of an identity-centric perimeter, at present this is largely limited to **DEVICE**-centric and **USER**-centric identities. However, there are many other identity attributes from the other asset classes that could be factored in as shown in Figure 48. Taking this approach could increase the identity management burden, but it may be appropriate in circumstances where higher levels of trustworthiness are needed for access.

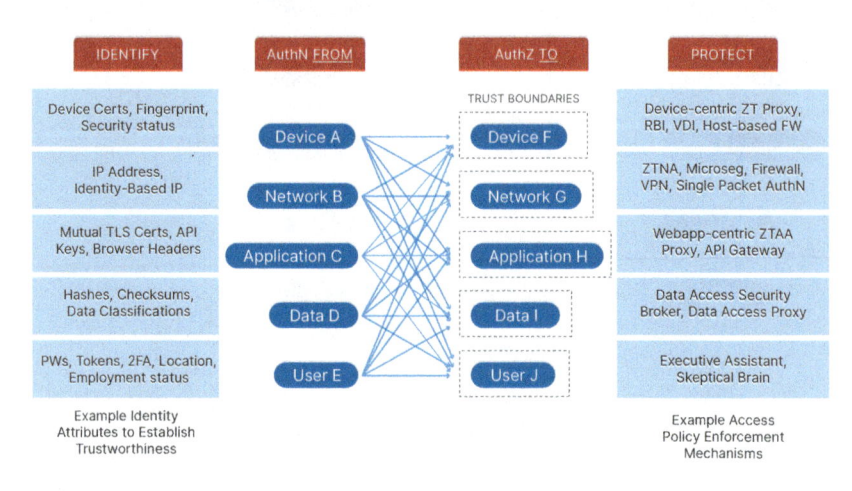

Figure 48: Consuming More Identity Attributes for Greater Trustworthiness

With these various forms of zero trust mapped to the Cyber Defense Matrix, as shown in Figure 49, the concept of zero trust can easily be applied across the board to other asset classes.

This gives a much broader range of options for how we can design towards zero trust access. This is an illustration of the continuing power of the Cyber Defense Matrix — through pattern matching and the imposition of consistency — to not just illuminate current security trends and thinking, but to also predict the emergence of new technologies and security approaches.

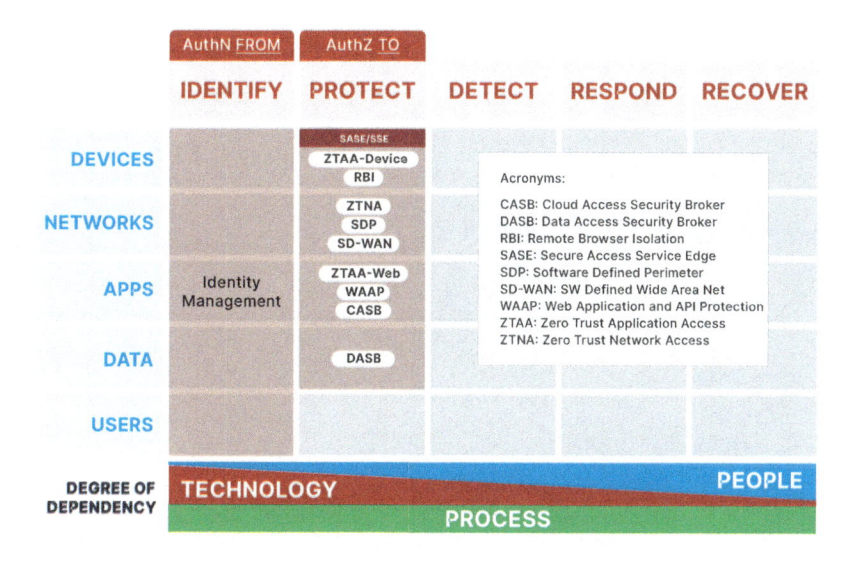

Figure 49: Mapping of Zero Trust Design Patterns

Extended Detection & Response and Managed Detection & Response

Attackers do not stay in designated lanes. As John Lambert once said, they think in graphs and leverage whatever assets are within reach to move towards their objective.[3] These assets are not strictly limited to one type of asset class. If attackers can pivot to a SaaS **APPLICATION** after compromising an endpoint **DEVICE** or establish an unmonitored rogue system on a **NETWORK**, then we need tooling that can also track an attacker across this broad range of assets. Hunting for lateral movement requires tools that enable lateral analysis across multiple asset classes.

Previously, we would find asset-specific hunting tools, such as endpoint/ **DEVICE** detection and response (EDR), **NETWORK** detection and response (NDR), and **USER** behavior analytics (UBA). The **NETWORK** detection capabilities originally emerged from an older category called Network Traffic Analysis (NTA), which typically consumed network flow logs to uncover malicious activity within the **NETWORK**. These capabilities would map into

---

3 https://git.io/fpfZ5

the Cyber Defense Matrix as shown in Figure 50.[4]

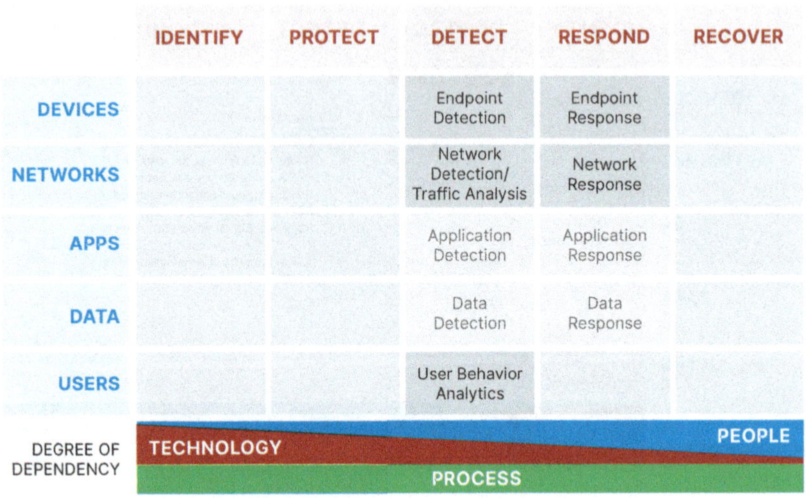

	IDENTIFY	PROTECT	DETECT	RESPOND	RECOVER
DEVICES			Endpoint Detection	Endpoint Response	
NETWORKS			Network Detection/ Traffic Analysis	Network Response	
APPS			Application Detection	Application Response	
DATA			Data Detection	Data Response	
USERS			User Behavior Analytics		

DEGREE OF DEPENDENCY — TECHNOLOGY ... PEOPLE / PROCESS

Figure 50: Mapping of Detection and Response Capabilities

However, these capabilities were operated independently and did not correlate suspicious events across these separate asset classes. The need to conduct this cross-asset analysis resulted in the capability known as User and Entity Behavior Analytics (UEBA). This capability combined detection capabilities across **DEVICES**, **NETWORKS**, and **USERS**. Subsequently, eXtended Detection and Response (XDR) emerged with the addition of telemetry from cloud assets (both IaaS/PaaS and SaaS) and the ability to drive response action through security orchestration and automated response (SOAR) systems as shown in Figure 51.[5]

---

4 As of the writing of this book, the capabilities of APPLICATION detection and response and DATA detection and response (dance party anyone?) are now just emerging in the market and have not gained significant traction, but the matrix easily demonstrates how these capabilities can be anticipated.

5 The ambiguity of the term "extended" should be evident. As mentioned previously, the Cyber Defense Matrix helps clarify what "extended" should mean and what we should be asking of vendors when they claim to have a complete solution that addresses the full range of assets.

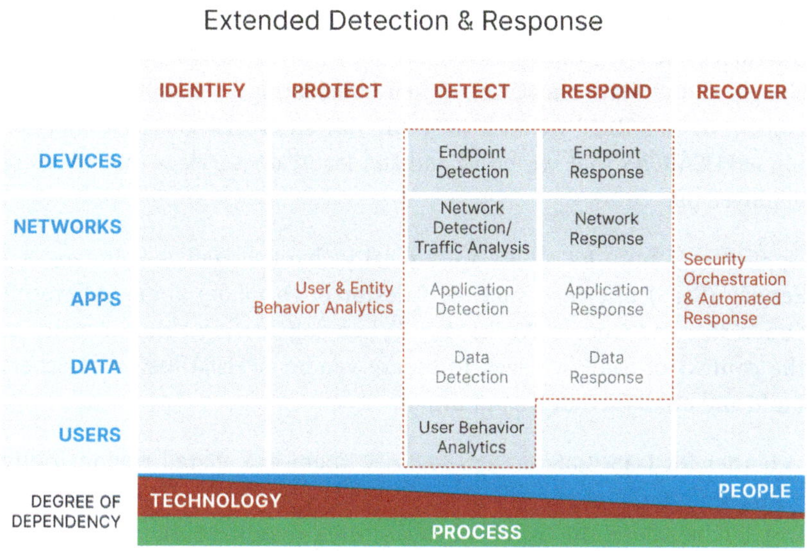

Extended Detection & Response

	IDENTIFY	PROTECT	DETECT	RESPOND	RECOVER
DEVICES			Endpoint Detection	Endpoint Response	
NETWORKS			Network Detection/ Traffic Analysis	Network Response	Security Orchestration & Automated Response
APPS		User & Entity Behavior Analytics	Application Detection	Application Response	
DATA			Data Detection	Data Response	
USERS			User Behavior Analytics		

DEGREE OF DEPENDENCY: TECHNOLOGY → PEOPLE / PROCESS

Figure 51: Convergence of UEBA and SOAR to Create XDR

As the Cyber Defense Matrix shows, as we move to the functions of **DETECT** and **RESPOND**, we have a greater dependency on *PEOPLE* to perform these functions. Despite the marketing language of vendors selling XDR solutions claiming that their products can displace *PEOPLE*, they simply do not operate effectively without knowledgeable and skilled personnel. However, many organizations have difficulty finding and hiring qualified personnel. As such, we have services such as Managed Detection & Response (MDR) that bring the necessary staff with the skillsets to address this dependency on *PEOPLE*.

Cloud Security

The term "cloud" is often as ambiguously defined as its physical namesake. It can refer to any number of assets, but in general, cloud security capabilities are easier to define and map to the Cyber Defense Matrix around two frames of reference:

- Securing applications that we build
- Securing applications built by others

Applications that we build typically leverage Infrastructure as a Service

(IaaS) and Platform as a Service (PaaS) offerings from major public cloud providers such as Amazon Web Services (AWS), Microsoft Azure, and Google Cloud Platform (GCP). When considering cloud security in the context of IaaS/PaaS, we seek to secure our enterprise and customer-facing **APPLICATIONS** that we build and the IaaS/PaaS services that they are deployed on.

Applications built by others are typically characterized as Software as a Service (SaaS) offerings such as Salesforce, Dropbox, Zoom, Microsoft Office 365, and Google Workspace. When considering cloud security in the context of SaaS, we seek to secure vendor **APPLICATIONS** that others build and our **DATA** that we put into it.

Whether for IaaS/PaaS or SaaS security, there is a shared responsibility with the underlying provider that limits what we can directly see and control. Furthermore, the technical capabilities, required skill sets, and governance processes are distinct to each type of cloud security. However, the macro pattern of what we want to accomplish is generally the same. We want to ensure that we can extend our security controls to perform these functions:

- **IDENTIFY:** The environment is catalogued and configured in accordance with best practices without unintended exposures.
- **PROTECT:** If there are any problems, we want to be able to fix them easily. We also want to ensure that the environment is properly used and properly accessed.
- **DETECT** and **RESPOND:** If an intrusion occurs in this environment, then we want capabilities to find and handle such a compromise.

The **DETECT** and **RESPOND** capabilities for "cloud" were discussed previously under the topic of XDR, so I will not repeat them here. However, the capabilities under **IDENTIFY** and **PROTECT** deserve special attention because they have spawned a whole new set of buzzwords that can be understood more easily when mapped to the Cyber Defense Matrix.

Cloud Security Posture Management and SaaS Security Posture Management

The most common reason for security breaches in the cloud is misconfiguration by the customer. It is then no surprise that we see a plethora of solutions addressing this common challenge. The security categories of Cloud Security Posture Management (CSPM) and SaaS Security Posture Management (SSPM) address this problem space. CSPM is sometimes characterized as addressing the SaaS part of the problem, but most often, it is exclusively focused on securing IaaS/PaaS configurations.[6] SSPM is exclusively focused on SaaS applications, but with deeper configuration support for a larger variety of SaaS applications.

For CSPM and SSPM, the main function is to find configuration flaws, which map to the function of **IDENTIFY**. Some of these tools purport to fix these flaws, in which case, they would also map to the function of **PROTECT**, but many CSPM and SSPM solutions do not actually fix the flaw but simply provide guidance on how to fix it. The table below shows how various cloud resources map to the asset classes of the Cyber Defense Matrix for CSPM and SSPM. The asset class mapping for a given CSPM or SSPM solution would depend upon the type of cloud resource being assessed. Not all CSPM solutions cover all types of cloud resources.

	IaaS/PaaS (CSPM)	SaaS (SSPM)
Devices	Containers, Compute, Hosts	N/A
Networks	VPCs, VPNs, CDNs, DNS	N/A
Applications	Microservices, Serverless Functions	e.g., Salesforce, Office365
Data	Datastores, Databases, Files	e.g., Box, Dropbox, Google Drive
Users	Accounts, User Roles	N/A

Table 6: Asset Coverage for CSPM and SSPM

---

6 There is a long tail of SaaS applications. If there is support for SaaS applications, it is usually a very limited set of the most popular ones.

Cloud Workload Protection Platform and SaaS Data Loss Prevention

In addition to ensuring that the cloud resource is configured properly, we also want to ensure that it is used properly. With IaaS/PaaS, this usually involves a technology capability called Cloud Workload Protection Platforms (CWPP). This capability covers part of the shared responsibility model that is entirely within the customer's area of responsibility, namely the content within a virtual machine, container, or serverless function that runs the application workload. Despite having the word "protection" in the name, some CWPP solutions only perform the **IDENTIFY** function of capturing vulnerabilities present in a workload. Others are able to orchestrate the removal of those vulnerabilities through the development pipeline or the build process, in which case, those CWPP capabilities would also map to the **PROTECT** function. The asset class mapping is a combination of **DEVICES** (for virtual machines) and **APPLICATIONS** (for containers and serverless functions).

For SaaS, the primary concerns related to usage revolve around the unmanaged or unintended proliferation of corporate data. As such, the primary capability to address this concern is SaaS Data Loss Prevention (SaaS DLP) capabilities. SaaS DLP features are available within the suite of capabilities of Cloud Access Security Brokers (CASB) products, but some products operate natively within the SaaS application through various API hooks offered by the SaaS provider. In other cases, the SaaS provider itself offers DLP as an add-on for higher level, enterprise-tier subscriptions. When it comes to the mapping of this category, DLP is certainly within the function of **PROTECT**. However, when it comes to the mapping of the asset, it may get a little confusing. The most intuitive mapping is to **DATA-PROTECT**. However, this would not be on the vendor layer of the matrix since we are not actually trying to **PROTECT** the vendor's **DATA** but our own.

Cloud Infrastructure Entitlement Management and SaaS Access Management

In addition to ensuring that our cloud resources are properly configured and used, we also want to ensure that access is properly managed. The product category of Cloud Infrastructure Entitlement Management (CIEM), also known as Cloud Identity Governance, helps us manage permissions and entitlements so that we can adhere to the principles of least privilege and limit the blast radius of any compromised accounts within IaaS and PaaS environments.

There are similar capabilities emerging in the SaaS space; however, because of the large variety of SaaS applications (as opposed to the three major IaaS/PaaS providers), the ability to manage entitlements and permissions across a broad range of SaaS applications is difficult for most vendors to support. As such, most vendors in this space cover only the most popular SaaS applications.

Both CIEM and SaaS Access Management solutions claim to cover both human and non-human identities, but the main strength of these solutions at the current time are in managing permissions and entitlements for human identities. As such, the capabilities as they exist today map to **USER-IDENTIFY**.

For IaaS/PaaS, these three categories of capabilities map to the Cyber Defense Matrix as shown in Figure 52.

Another new category has emerged called **Cloud Native Application Protection Platforms** (CNAPP) that combine CSPM, CWPP, and CIEM capabilities. This makes sense with respect to the natural adjacencies among these capabilities. There has not been an equivalent convergence in the SaaS security landscape, but one may emerge eventually.

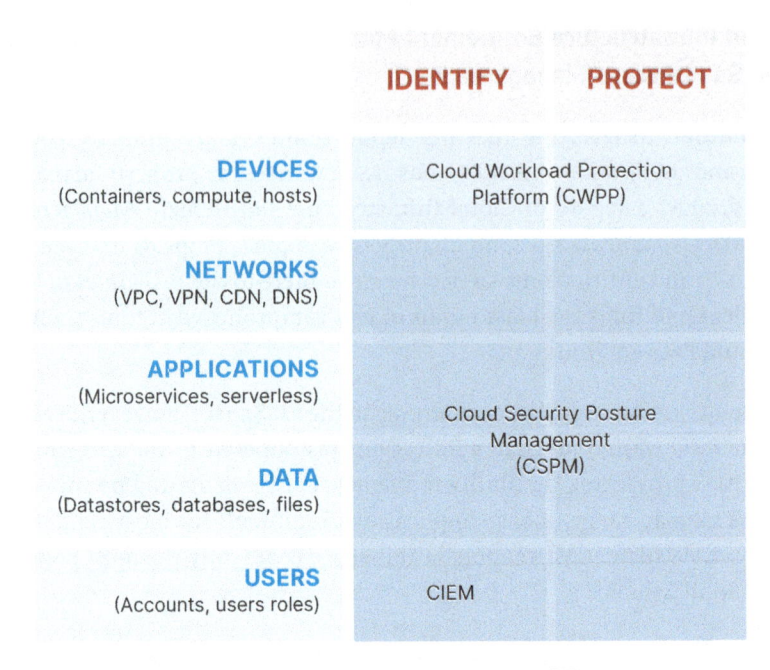

	IDENTIFY	PROTECT
DEVICES (Containers, compute, hosts)	Cloud Workload Protection Platform (CWPP)	
NETWORKS (VPC, VPN, CDN, DNS)		
APPLICATIONS (Microservices, serverless)	Cloud Security Posture Management (CSPM)	
DATA (Datastores, databases, files)		
USERS (Accounts, users roles)	CIEM	

Figure 52: Mapping of CSPM, CWPP, and CIEM

Cyber Asset Attack Surface Management

Attack surface management (ASM) enables us to gain structural awareness of our environment, particularly as it pertains to what is exposed to attackers. This awareness is typically gained through both external scanning techniques and internal data aggregation, often through API integrations. These two approaches complement each other well to validate the findings from both sources. Ideally, the attack surfaces discovered through external scanning will match that which is known through internal data sources.

The discovery, prioritization, and monitoring of attack surfaces align under **IDENTIFY**. Fixing any issues would align under **PROTECT**; however most ASM capabilities at the present time only perform the **IDENTIFY**-related activities.

The asset classes associated with ASM vary based on the vendor and their specific techniques for discovering assets and their associated attack surfaces, but most focus on **DEVICES**, **NETWORKS**, and **APPLICATIONS**. The recently defined category of Cyber Asset Attack Surface Management (CAASM) takes a broader view of what constitutes an asset and counts everything under an organization's control. However, without a prescriptive list, we may leave out an asset for consideration. This is where the Cyber Defense Matrix can come to the rescue again to provide a comprehensive list of assets that we care about. These include the five assets directly owned by the organization, but also assets owned by our vendors, customers, and employees as discussed in Chapter 1.

Moreover, the attack surfaces associated with these assets should not be seen in isolation. They are connected together, but often in ways that we cannot easily understand. Capabilities in the CAASM category, done properly, should aggregate information about all the assets defined in the multi-dimensional Cyber Defense Matrix and connect them together in a graph to enable us to understand how the assets and their associated attack surfaces relate to one another.

Data Security Posture Management / Cloud DLP

Data security posture management (DSPM) provides capabilities to discover, catalog, and classify the content of data stores and file repositories within and organization's environment. As such, DSPM maps to **DATA-IDENTIFY**. This information is used to provide context about what should be the proper configuration of those resources.

The emergence of the DSPM category should raise further questions around the meaning of "Security Posture Management" as a term. *Posture* is a configuration that is not inherently good or bad. A person standing up straight would be in a bad configuration if they wanted to sleep and a person lying prone would be in a bad configuration if they wanted to stay awake. Whether a configuration is actually good or bad depends on having more context. In contrast, a vulnerability is inherently bad regardless of context.[7]

---

[7] Whether one choosed to fix or address the vulnerability requires additional context (e.g., threat and impact), but this does not change the fact that a vulnerability is still bad regardless of context.

Because bad configurations are context-dependent, it is important that we get as much context as needed to ensure that we make the right decisions on whether or not something that is flagged as a misconfiguration is actually bad. We often get reports from other tools (e.g., CSPM) about potentially misconfigured resources, such as an open S3 bucket or unencrypted datastore. However, these reports are often devoid of context, requiring the gathering of more information about the nature of the content or the purpose of the resource to properly determine if the current configuration is actually bad.

DSPM products gather additional context based on the actual content of the data, observed access patterns, and existing permissions to help determine whether or not the existing configuration is actually bad. Because DSPM tools require a peek into one's data, including potentially sensitive content, many organizations will likely choose to run the scanning component of a DSPM capability within the boundaries of their direct control and feed only the metadata of the content (e.g.., the data classification) back to the DSPM vendor.

The DSPM category is gradually converging with the DLP category with a specific focus on IaaS- and SaaS-based data repositories (i.e., Cloud DLP). If DSPM incorporates measures to prevent actual data movement, then it may eventually also map to **DATA-PROTECT**.

CHAPTER 10

Conclusion

*I may not have gone where I intended to go,
but I think I have ended up where I needed to be.*
-Douglas Adams

The Road Ahead

As we complete our journey together —or perhaps, just begin it — I hope that my navigational guide has been helpful, and that you have discovered possibilities for putting the Cyber Defense Matrix to use in your own environment.[1] As you proceed, I encourage you to start with something simple. At its core, the Cyber Defense Matrix is a simple taxonomy for organizing cybersecurity-relevant information. As you encounter different aspects of the cybersecurity landscape over the course of a typical day, or in a security conference expo hall, consider taking one or two of them and determine where they fit on the matrix, and why. It is a good exercise for strengthening your grasp of the concepts underlying the Cyber Defense Matrix, and good practice in applying it for the use cases you have identified for your organization.

As you go about your journey, here are a few tips that will keep you on the straight and narrow path:

1. Try to adhere to the principle of sticking something in one box and only one box. This is a forcing function that will cause you to think harder about the real nature of the capability or activity that you are

---

1 As the creator of the Cyber Defense Matrix, critics would be right to point out that I will have a tendency to see everything as a nail when I am wielding a hammer. I generally think that I am not wielding just a hammer, but rather a toolkit with several (but still limited) tools at my disposal.

weighing. The end conclusion may be that it fits in multiple boxes, but that should be the rare exception, not the rule. If a particular capability or vendor product still seems to fit in multiple boxes, then see if it is actually a bundled capability that can be functionally decomposed into separate boxes.

2. Avoid the temptation to add another row or column to the matrix. Instead, consider adding a descriptor to the asset class. Or add another dimension — but be mindful of the next recommendation if you go that route.

3. Be wary of adding new dimensions. It can be hard enough to understand just the two dimensional view. Three or more dimensions would not only be harder to represent on paper; they would generally cause most heads to explode.[2] Still, adding another dimension may be unavoidable at times to properly represent the variety of capabilities that we see in cybersecurity. If and when you do add another dimension, be mindful that both the horizontal and vertical axes need to be applied.

4. Remember that the Cyber Defense Matrix is most useful when seen from a strategic standpoint. There is only so much detail that can be shown on a two dimensional matrix before it gets too confusing. If you find yourself desiring more detail than the Cyber Defense Matrix can offer, you are likely moving into more tactical territory and may want to consider other frameworks that are complementary to the Cyber Defense Matrix.

The use cases I have discussed in this book are only a few of the dozens I have given thought to — with unlimited time and opportunity, I could have continued at virtually unlimited length! Perhaps it is best that I have presented the Cyber Defense Matrix in more compact form for the time being. I may one day return to document additional use cases, but in the meantime, I would love to hear from you and your fellow practitioners about the use cases you have explored, the utility you have found, and your experiences working with it.

---

[2] I have a few more complex use cases that require the matrix to be seen from several dimensions but I think that would be more appropriate for future volumes or for those willing to have their heads hurt a little.

Ultimately, my intention in creating the Cyber Defense Matrix has been to bring clarity and insight to an often excessively confusing cybersecurity landscape. I do not mean to cast aspersions on the vendors and marketers whose own terminology and classifications at time contribute to this obscurity. In a world full of passionate people working hard to stem the tide of cyberthreats, it is understandable that we end up with conflicting versions of reality: which priorities matter most, how solution categories should be defined and named, and whose value proposition stands supreme. Hopefully, the structure, discipline, and consistency offered by the matrix will help practitioners and tool vendors work together more productively to achieve optimal cyber defense for each customer — to ensure comprehensive protection for their assets, to respond effectively to incidents, to close existing gaps and prevent new ones from arising, and to create a coherent and efficient strategic roadmap moving forward.

I thank you for the interest you have taken in the Cyber Defense Matrix and the time you have invested to understand its purpose and function. I hope it serves you well!

Acknowledgements

It has been an arduous journey writing this book, and I would not have made it this far without the assistance of many others who walked alongside me.

I am particularly thankful to my wife, Gracie, and kids, Caleb, Jason, Kristi, and Renee, whose patience, encouragement, and love kept me going as I toiled away putting my thoughts to paper. Their many hands made for light work as they helped prepare sticky notes for the several workshops that I held at RSA. During the COVID-19 pandemic, they heard me talk over Zoom about many of the concepts written here. If I had them take notes or recount back to me what I said over Zoom, I probably would have gotten this book done faster. Maybe for the next edition…

I am also thankful to my work colleagues across three different organizations who supported me through feedback, encouragement, and resources. These include Erkang Zheng, Tyler Shields, and Melissa Pereira at JupiterOne for helping me get the book over the finish line; Yoav Leitersdorf and John Brennan at YL Ventures for giving me the opportunity to test my use cases on new entrepreneurs, and even making investment decisions based on my ideas; and Craig Froelich, Christofer Hoff, and Robert Brown at Bank of America for giving me the latitude and encouragement to bring this idea into the public forum.

I would also like to thank the many friends and supporters of the Cyber Defense Matrix. I am blessed that there are so many to list, and if I left you off by mistake, please forgive me.

- Wendy Nather for being such a fierce advocate for the Cyber Defense Matrix and weaving it into so many of her talks
- Ron Gula for prompting me to dream bigger than I would have considered
- Bryan Ware for giving me the opportunity to test the Cyber Defense Matrix at a much larger scale
- Anthony Johnson for helping me ideate on use cases over Korean BBQ
- Josh Thurston for battle-testing the methodology against many edge cases

- Shaun Waterman for helping me refine my scattered thoughts into something more coherent
- Will Lin for helping me see how the Cyber Defense Matrix could be used by venture capitalists
- Allan Alford, Ryan Bowling, Christophe Foulon, Paul Ihme, Tommy Jinks, Julia Knecht, Erich Smith, Ray Winder, Andrea Weisberger, and Anne Marie Zettlemoyer for being my faithful facilitators at various RSA workshops
- Jon Hawes, Guillaume Ross, Adrian Sanabria, and Martha VanDriel for being my co-speakers at those workshops, sharing their personal use cases of the Cyber Defense Matrix
- Gadi Evron for giving me the forum at A Conference for Defense (ACoD) to further develop these ideas among a larger community
- Andrew Morgan for helping me understand how the Cyber Defense Matrix could be applied for managed service providers
- Matt Zanderigo for giving me a stage within AWS Marketplace to share the Cyber Defense Matrix to a massive audience
- Ross Young for his detailed feedback and adjacent ideas that have made our respective models even better

About the Author

Sounil Yu is a security innovator with a deep knowledge of computer systems and a career spanning over three decades as an executive leader of information technology and security in the federal government, military, and private sectors. Currently, he is the CISO and Head of Research at JupiterOne, a fast-growing cybersecurity startup building a cloud-native asset management and security platform.

Prior to joining JupiterOne, he was the CISO-in-Residence at YL Ventures, a venture capital firm focused on early-stage Israeli cybersecurity startups. Before that, he was the Chief Security Scientist at Bank of America, leading a cross-functional team driving security innovation by leveraging unconventional thinking and alternative approaches to address key challenges in security. Examples of these innovative approaches include the creation of the Cyber Defense Matrix and the DIE Triad, which have shaped the views of the industry, regulators, and the overall security ecosystem.

Earlier in his career, he helped improve information security at several institutions spanning from Fortune 100 companies with three letters on the stock exchange to secretive three-letter agencies that are not.

Sounil serves on the Board of Advisors for the FAIR Institute and Project N95. He guest lectures at Carnegie Mellon University and was an adjunct professor at George Mason University and Yeshiva University's Katz School of Science and Health, teaching the fundamentals of Cybersecurity Technologies. He also is a fellow at GMU Scalia Law School's National Security Institute and advises several startups. In addition, he co-chairs Art into Science: A Conference on Defense and previously served as co-chair of the OASIS OpenC2 Standard.

He has received numerous awards, including the 2022 Lifetime Achievement Difference Maker Award by the SANS Institute, 2021 Influencer of the Year by SC Awards, 2021 Top 10 CISO by Black Unicorn Awards, and 2020 Most Influential People in Security by *Security Magazine*. Widely regarded as the go-to source on cybersecurity trends, Sounil is quoted and published in media such as *The Wall Street Journal*, *Forbes*, *ZDNet*, *Dark Reading*, *ThreatPost*, *Security Boulevard*, and *SC Magazine*.

Sounil holds a CISSP certification and earned a master's degree in Electrical Engineering from Virginia Tech and multiple bachelor's degrees in Electrical Engineering and Economics from Duke University. He has over 20 granted patents covering a wide range of topics, including threat modeling, graph databases, intrusion deception, endpoint security monitoring, tracking media leaks, attributing malicious requests, attributing devices to organizations, detecting logic bombs, security portfolio optimization, and neutralizing stolen files. He is a frequently sought speaker at major security conferences, including RSA, DEFCON, SANS, KubeCon, and FS-ISAC.

Sounil lives with his family in Virginia, where he enjoys reading, playing board games, and taking naps, not necessarily in that order.

CYBER DEFENSE MATRIX

	IDENTIFY	PROTECT	DETECT	RESPOND	RECOVER
DEVICES					
NETWORKS					
APPS					
DATA					
USERS					

DEGREE OF DEPENDENCY: TECHNOLOGY → PROCESS → PEOPLE

	IDENTIFY	PROTECT	DETECT	RESPOND	RECOVER
DEVICES	Asset Mgt, Vuln Scanner, Vuln Mgt, Certificate Mgt	AV, EPP, FIM, HIPS, Allowlisting, Vuln Mgt	Endpoint Detection, UEBA, XDR	EP Response, EP Forensics, SOAR	
NETWORKS	Netflow, Network Vuln Scanner	FW, IPS/IDS, Microseg, ESG, SWG, ZTNA	DDoS Detection, Net Traf Analysis, UEBA, XDR	DDoS Response, NW Forensics, SOAR	
APPS	SAST, DAST, SW Asset Mgt, Fuzzers	RASP, WAF, ZT Access Proxy	Src Code Compromise, Logic Bomb, App IDS, ADR		
DATA	Data Audit, Discovery, Classification	Encryption, DLP, Tokenization, DRM, DBAM, DB Proxy	Deep Web Analysis, Data Leak Discovery, DDR	DRM, Breach Response	Backup
USERS	Phishing Sim, Background Chk, MFA	Security Training & Awareness	Insider Threat, User Behavior Analytics, XDR		

DEGREE OF DEPENDENCY: TECHNOLOGY — PROCESS — PEOPLE

Made in the USA
Monee, IL
16 April 2023